T0328907

The Cellular Secretome and Organ Crosstalk

The Cellular Secretome and Organ Crosstalk

Juergen Eckel

Academic Press is an imprint of Elsevier
125 London Wall, London EC2Y 5AS, United Kingdom
525 B Street, Suite 1650, San Diego, CA 92101, United States
50 Hampshire Street, 5th Floor, Cambridge, MA 02139, United States
The Boulevard, Langford Lane, Kidlington, Oxford OX5 1GB, United Kingdom

Notices
Knowledge and best practice in this field are constantly changing. As new research and experience broaden our understanding, changes in research methods, professional practices, or medical treatment may become necessary.

Practitioners and researchers must always rely on their own experience and knowledge in evaluating and using any information, methods, compounds, or experiments described herein. In using such information or methods they should be mindful of their own safety and the safety of others, including parties for whom they have a professional responsibility.

To the fullest extent of the law, neither the Publisher nor the authors, contributors, or editors, assume any liability for any injury and/or damage to persons or property as a matter of products liability, negligence or otherwise, or from any use or operation of any methods, products, instructions, or ideas contained in the material herein.

Library of Congress Cataloging-in-Publication Data
A catalog record for this book is available from the Library of Congress

British Library Cataloguing-in-Publication Data
A catalogue record for this book is available from the British Library

ISBN: 978-0-12-809518-8

For information on all Academic Press publications visit our website at
https://www.elsevier.com/books-and-journals

Working together
to grow libraries in
developing countries

www.elsevier.com • www.bookaid.org

Publisher: John Fedor
Acquisition Editor: John Fedor
Editorial Project Manager: Carlos Rodriguez
Production Project Manager: Punithavathy Govindaradjane
Cover Designer: Matthew Limbert

Typeset by TNQ Technologies

Contents

About the Author

Juergen Eckel is a Professor of Clinical Biochemistry, and he has been working at the German Diabetes Center in Düsseldorf since 1978. His field of study includes insulin signaling, insulin resistance, type 2 diabetes, obesity, adipose tissue and skeletal muscle biology, and organ crosstalk. His research group was the first to directly demonstrate the communication between human adipocytes and skeletal muscle cells. From 2006 to 2011, Prof. Eckel worked as the Acting Director of the Institute of Clinical Biochemistry and Pathobiochemistry at the German Diabetes Center. From 2011 to 2016, he directed the Paul-Langerhans-Group for Integrative Physiology.

Prof. Eckel served as a chairperson for several EU COST Actions between 2000 and 2011 and coordinated the EU FP7 project, ADAPT. He received several awards and published more than 200 original papers and reviews.

General Introduction

1

CHAPTER OUTLINE

Given the complexity of higher organisms, a sophisticated system of cell—cell and organ—organ communication is required to adapt and harmonize the multiple physiological functions of different organs and to orchestrate and regulate energy intake and metabolic performance. This system of biological communication, often called "organ crosstalk," makes it possible for the cells in one tissue to send information to cells in another tissue even at relative long distances. This process involves molecular, cellular, and neural factors and is a key determinant of physiological homeostasis. It is known for longtime that derangement in this organ—organ communication system can finally initiate multiorgan dysfunction (Virzi et al., 2014; Molls and Rabb, 2004). As a paradigm, the so-called cardiorenal syndrome is a well-known example where pathological changes in one organ can severely affect function in another organ. The cardiorenal syndrome is a complex entity including vessel inflammation, atherosclerosis, cardiac fibrosis, and hypertrophy (Kingma et al., 2017). Importantly, the biochemical abnormalities can adversely affect both cardiovascular and renal function (Feltes et al., 2008; Li et al., 2009).

It is now becoming more and more evident that organ crosstalk plays an unprecedented role in a variety of divergent and fundamental processes of physiological and pathophysiological importance. Thus, it was recently reported that the microbiome and the immune system are emerging as important players in regulating β-cell function and mass (Shirakawa et al., 2017). The rich innervation of islet cells indicates it is a prime organ for regulation by the nervous system. The authors discussed the potential implications of signals from these organ systems as well as those from bone, placenta, kidney, thyroid, endothelial cells, reproductive organs, and adrenal and pituitary glands that can directly affect β-cell biology. This ability of pancreatic β-cells to crosstalk with multiple nonmetabolic tissues points to a new avenue for improving β-cell function and/or mass in diabetic patients. Organ crosstalk may also underlie the well-known association of obesity and cancer. As recently reported, epidemiologic, clinical, and preclinical data suggest that within the growth-promoting, proinflammatory microenvironment, accompanying obesity,

The Cellular Secretome and Organ Crosstalk. https://doi.org/10.1016/B978-0-12-809518-8.00001-5

crosstalk between adipose tissue (comprised of adipocytes, macrophages, and other cells) and cancer-prone cells may occur via obesity-associated hormones, cytokines, and other mediators that have been linked to increased cancer risk and/or progression (Himbert et al., 2017). These authors suggest an organ-dependent crosstalk between adipose tissue and carcinomas via VEGF, IL6, TNFα, and other mechanisms. Moreover, visceral white adipose tissue plays a more central role, as it is more bioenergetically active and is associated with a more procancer secretome than subcutaneous adipose tissue. Finally, alcoholic liver disease most likely represents a paradigm of multiorgan crosstalk involving interaction of the liver with the adipose tissue, gut, brain, and lung (Poole et al., 2017).

In addition to interorgan crosstalk, intraorgan crosstalk has also gained considerable interest, and this paracrine communication within tissues plays multiple roles for immune modulation, differentiation, cellular growth and development, and metabolic regulation (Martin-Saavedra et al., 2017; Kim et al., 2018; Peteranderl et al., 2016; Bassino et al., 2015; Nitta and Orlando, 2013; Figeac et al., 2014; Ben-Batalla et al., 2013; Grubisha and DeFranco, 2013; Nov et al., 2013; Clarkin et al., 2008; Dietze et al., 2004). Given the complexity of organ crosstalk, it is not surprising that an array of mechanisms exists to mediate the communication process. At present, we only partly understand the language of different cells and tissues, and it is certainly a major challenge for future research to identify common signals or codes that may help to develop a unifying concept of organ crosstalk.

At present, we are facing a complex scenario of crosstalk mediators involving nutrients and metabolites, extracellular vesicles, and peptides and proteins that collectively may be called "organokines" (Choi, 2016). It is presently not clear whether these molecules fit into a hierarchy of communication mediators or if these molecules act independently at different levels of organ crosstalk. They may also have a time- and spatial-dependent relevance, and this will be an important issue for future research. It is well known that the flux of substrates, such as carbohydrates and lipids, through the liver can affect distal organs by means of their energy-sensing systems (Poole et al., 2017). Nutrients are also important players for the crosstalk between distal organs such as the liver and the central nervous system. In addition to nutrients, metabolites are also active players in crosstalk processes. However, this is an extremely complex setting, as the metabolome currently comprises about 40,000 experimentally detected and biologically expected human metabolites (Wishart et al., 2013). The precise role of the vast majority of these molecules in mediating cell-to-cell and/or organ-to-organ communication remains elusive. A major problem with the metabolome is the huge range of concentration and the heterogeneous structure of the constituting small molecules. This has hampered the systematic and detailed studies on the role of the metabolome in organ crosstalk, although it can be anticipated that these molecules play an important role in this process.

Very recent discoveries of novel mediators of organ communication relate to so-called extracellular vesicles that include microvesicles and exosomes (Whitham et al., 2018; Robado de Lope et al., 2018; Collison, 2018; Harisa et al., 2017; Lindoso et al., 2017; Al-Samadi et al., 2017). This novel principle of organ crosstalk may also play an important role in the pathogenesis of several diseases (Maji et al., 2017a,b; Huang-Doran et al., 2017). Extracellular vesicles is the collective term that includes exosomes, microvesicles (nanoparticles), and apoptotic bodies. Extracellular vesicles carry a diverse bioactive cargo of proteins, lipids, metabolites, DNA, and RNA (mRNA and small regulatory RNAs). These vesicles are released to the local tissues and the circulation in a continuous process that has been extensively investigated (Katzmann et al., 2001). Substantial evidence supports the notion that extracellular vesicles participate in organ crosstalk and function as conveyers of cellular information owing to their specific cargo (Valadi et al., 2007; Montecalvo et al., 2012). Application of transcriptomic, lipidomic, and proteomic technologies for analysis of extracellular vesicles from different cell types has resulted in a comprehensive view on the protein, lipid, and nucleic acid content of these vesicles (Keerthikumar et al., 2016). It is generally thought that exosomes can interact with a target cell, deliver its cargo to the cytosol of the target cell, and modulate its function by delivery of protein, translatable mRNA, and silencing of recipient genes by siRNA. Evidence exists that adipose-derived exosomes may play a role in the metabolic consequences of obesity and adipose tissue dysfunction (Deng et al., 2009). Extracellular vesicles have also been recognized to be implicated in a spectrum of different liver diseases including hepatocellular carcinoma, nonalcoholic fatty liver disease, and alcoholic liver disease (Hirsova et al., 2016). Despite the significant progress regarding the role of extracellular vesicles in organ communication, the key question, whether they are causal players or simply passive bystanders, remains unanswered. This is paralleled by many additional questions related to the specificity, kinetics, and characterization of extracellular vesicles, and specifically, the question related to their in vivo action profile. Future work on this very interesting topic will augment our picture of organ communication and certainly open new avenues for a targeted manipulation of dysregulated organ crosstalk.

At present, the most advanced understanding of cellular communication and organ crosstalk relates to organokines, which collectively describe peptides and proteins released from cells, and thus comprising the secreted proteome (Zhang et al., 2018; Shirakawa et al., 2017; Mitaka et al., 2014; Romacho et al., 2014). Although recent secretome analysis data show that most likely thousands of peptides and proteins are released by adipocytes, myocytes, liver cells, immune cells, and many others, they are a much more homogeneous class of molecules compared with the metabolome. Many of these molecules are cytokines including interleukins, interferons, chemokines, and similar molecules playing a role in immunoregulatory processes with either proinflammatory or antiinflammatory functions. In fact, the seminal work published by Spiegelman et al. about 25 years ago identified the cytokine TNFα as a critical signal mediating the negative crosstalk between adipose tissue and skeletal muscle (Hotamisligil et al., 1993, 1994). These findings set the stage for a tremendous and still-ongoing interest in organ crosstalk as a major element of metabolic homeostasis and a key driver for metabolic diseases. Given that most peptides and proteins engage cellular receptors and induce complex downstream signaling pathways that finally modify effector systems and affect the cellular phenotype, the organokine crosstalk from one cell to another could be successfully pinpointed to a mechanistic background in many cases. In this scenario, adipokines and myokines are the paradigm molecules that have been extensively analyzed in recent years (Chung and Choi, 2017; Oh et al., 2016; Duzova et al., 2018; Li et al., 2017; Rodriguez et al., 2017; Roh et al., 2016; Dalamaga, 2013; Yang et al., 2013; Granados et al., 2011; Trayhurn et al., 2011; Walsh, 2009). This has definitely led to a new understanding of metabolic regulation and homeostasis and has specifically unraveled new pathways leading to metabolic diseases such as insulin resistance and type 2 diabetes.

The overarching goal of this book is to provide a comprehensive look on the cellular secretome of adipocytes and myocytes and to put it in the functional context of organ crosstalk. This is described in detail in Chapters 2 and 3. Importantly, there is an interesting and still incompletely understood overlap between these molecules, leading to the term adipomyokines (Trayhurn et al., 2011). This topic is presented in Chapter 4. Finally, a dysregulated secretion appears to be of key importance for the development of a variety of metabolic diseases. Chapter 5 is focused on this interesting topic of high translational value. Finally, a Technical Annex (Chapter 6) is included describing an extensive collection of methods that have been used in the author's laboratory for many years. Many parts of this book are based on an extensive series of review papers on the topic of organ crosstalk prepared by the author and many coworkers at the German Diabetes Center in Düsseldorf, Germany, during the last 20 years. The author is deeply grateful for all these contributions that were instrumental to writing this book on organ crosstalk.

REFERENCES

Al-Samadi, A., Awad, S.A., Tuomainen, K., Zhao, Y., Salem, A., Parikka, M., Salo, T., 2017. Crosstalk between tongue carcinoma cells, extracellular vesicles, and immune cells in in vitro and in vivo models. Oncotarget 8, 60123—60134.

Bassino, E., Gasparri, F., Giannini, V., Munaron, L., 2015. Paracrine crosstalk between human hair follicle dermal papilla cells and microvascular endothelial cells. Exp. Dermatol. 24, 388—390.

Ben-Batalla, I., Schultze, A., Wroblewski, M., Erdmann, R., Heuser, M., Waizenegger, J.S., Riecken, K., Binder, M., Schewe, D., Sawall, S., Witzke, V., Cubas-Cordova, M., Janning, M., Wellbrock, J., Fehse, B., Hagel, C., Krauter, J., Ganser, A., Lorens, J.B., Fiedler, W., Carmeliet, P., Pantel, K., Bokemeyer, C., Loges, S., 2013. Axl, a prognostic and therapeutic target in acute myeloid leukemia mediates paracrine crosstalk of leukemia cells with bone marrow stroma. Blood 122, 2443—2452.

Choi, K.M., 2016. The impact of organokines on insulin resistance, inflammation, and atherosclerosis. Endocrinol. Metab. (Seoul) 31, 1—6.

Chung, H.S., Choi, K.M., 2017. Adipokines and myokines: a pivotal role in metabolic and cardiovascular disorders. Curr. Med. Chem. https://doi.org/10.2174/09298673256661712 05144627.

Clarkin, C.E., Emery, R.J., Pitsillides, A.A., Wheeler-Jones, C.P., 2008. Evaluation of VEGF-mediated signaling in primary human cells reveals a paracrine action for VEGF in osteoblast-mediated crosstalk to endothelial cells. J. Cell. Physiol. 214, 537—544.

Collison, J., 2018. Bone: extracellular vesicles in bone cell crosstalk. Nat. Rev. Rheumatol. 14, 2–3.

Dalamaga, M., 2013. Interplay of adipokines and myokines in cancer pathophysiology: emerging therapeutic implications. World J. Exp. Med. 3, 26–33.

Deng, Z.B., Poliakov, A., Hardy, R.W., Clements, R., Liu, C., Liu, Y., Wang, J., Xiang, X., Zhang, S., Zhuang, X., Shah, S.V., Sun, D., Michalek, S., Grizzle, W.E., Garvey, T., Mobley, J., Zhang, H.G., 2009. Adipose tissue exosome-like vesicles mediate activation of macrophage-induced insulin resistance. Diabetes 58, 2498–2505.

Dietze, D., Ramrath, S., Ritzeler, O., Tennagels, N., Hauner, H., Eckel, J., 2004. Inhibitor kappaB kinase is involved in the paracrine crosstalk between human fat and muscle cells. Int. J. Obes. Relat. Metab. Disord. 28, 985–992.

Duzova, H., Gullu, E., Cicek, G., Koksal, B.K., Kayhan, B., Gullu, A., Sahin, I., 2018. The effect of exercise induced weight-loss on myokines and adipokines in overweight sedentary females: steps-aerobics vs. jogging-walking exercises. J. Sports Med. Phys. Fit. 58, 295–308.

Feltes, C.M., Van Eyk, J., Rabb, H., 2008. Distant-organ changes after acute kidney injury. Nephron. Physiol. 109, p80–p84.

Figeac, F., Lesault, P.F., Le Coz, O., Damy, T., Souktani, R., Trebeau, C., Schmitt, A., Ribot, J., Mounier, R., Guguin, A., Manier, C., Surenaud, M., Hittinger, L., Dubois-Rande, J.L., Rodriguez, A.M., 2014. Nanotubular crosstalk with distressed cardiomyocytes stimulates the paracrine repair function of mesenchymal stem cells. Stem Cell. 32, 216–230.

Granados, N., Amengual, J., Ribot, J., Palou, A., Bonet, M.L., 2011. Distinct effects of oleic acid and its trans-isomer elaidic acid on the expression of myokines and adipokines in cell models. Br. J. Nutr. 105, 1226–1234.

Grubisha, M.J., DeFranco, D.B., 2013. Local endocrine, paracrine and redox signaling networks impact estrogen and androgen crosstalk in the prostate cancer microenvironment. Steroids 78, 538–541.

Harisa, G.I., Badran, M.M., Alanazi, F.K., Attia, S.M., 2017. Crosstalk of nanosystems induced extracellular vesicles as promising tools in biomedical applications. J. Membr. Biol. 250, 605–616.

Himbert, C., Delphan, M., Scherer, D., Bowers, L.W., Hursting, S., Ulrich, C.M., 2017. Signals from the adipose microenvironment and the obesity-cancer link-a systematic review. Canc. Prev. Res. 10, 494–506.

Hirsova, P., Ibrahim, S.H., Verma, V.K., Morton, L.A., Shah, V.H., LaRusso, N.F., Gores, G.J., Malhi, H., 2016. Extracellular vesicles in liver pathobiology: small particles with big impact. Hepatology 64, 2219–2233.

Hotamisligil, G.S., Murray, D.L., Choy, L.N., Spiegelman, B.M., 1994. Tumor necrosis factor alpha inhibits signaling from the insulin receptor. Proc. Natl. Acad. Sci. U. S. A. 91, 4854–4858.

Hotamisligil, G.S., Shargill, N.S., Spiegelman, B.M., 1993. Adipose expression of tumor necrosis factor-alpha: direct role in obesity-linked insulin resistance. Science 259, 87–91.

Huang-Doran, I., Zhang, C.Y., Vidal-Puig, A., 2017. Extracellular vesicles: novel mediators of cell communication in metabolic disease. Trends Endocrinol. Metabol. 28, 3–18.

Katzmann, D.J., Babst, M., Emr, S.D., 2001. Ubiquitin-dependent sorting into the multivesicular body pathway requires the function of a conserved endosomal protein sorting complex, ESCRT-I. Cell 106, 145–155.

Keerthikumar, S., Chisanga, D., Ariyaratne, D., Al Saffar, H., Anand, S., Zhao, K., Samuel, M., Pathan, M., Jois, M., Chilamkurti, N., Gangoda, L., Mathivanan, S., 2016. ExoCarta: a web-based compendium of exosomal cargo. J. Mol. Biol. 428, 688−692.

Kim, M., Lee, J., Park, T.J., Kang, H.Y., 2018. Paracrine crosstalk between endothelial cells and melanocytes through clusterin to inhibit pigmentation. Exp. Dermatol. 27, 98−100.

Kingma, J.G., Simard, D., Rouleau, J.R., Drolet, B., Simard, C., 2017. The physiopathology of cardiorenal syndrome: a review of the potential contributions of inflammation. J. Cardiovasc. Dev. Dis. 4.

Li, F., Li, Y., Duan, Y., Hu, C.A., Tang, Y., Yin, Y., 2017. Myokines and adipokines: involvement in the crosstalk between skeletal muscle and adipose tissue. Cytokine Growth Factor Rev. 33, 73−82.

Li, X., Hassoun, H.T., Santora, R., Rabb, H., 2009. Organ crosstalk: the role of the kidney. Curr. Opin. Crit. Care 15, 481−487.

Lindoso, R.S., Collino, F., Vieyra, A., 2017. Extracellular vesicles as regulators of tumor fate: crosstalk among cancer stem cells, tumor cells and mesenchymal stem cells. Stem Cell Invest. 4, 75.

Maji, S., Matsuda, A., Yan, I.K., Parasramka, M., Patel, T., 2017a. Extracellular vesicles in liver diseases. Am. J. Physiol. Gastrointest. Liver Physiol. 312, G194−G200.

Maji, S., Yan, I.K., Parasramka, M., Mohankumar, S., Matsuda, A., Patel, T., 2017b. In vitro toxicology studies of extracellular vesicles. J. Appl. Toxicol. 37, 310−318.

Martin-Saavedra, F., Crespo, L., Escudero-Duch, C., Saldana, L., Gomez-Barrena, E., Vilaboa, N., 2017. Substrate microarchitecture shapes the paracrine crosstalk of stem cells with endothelial cells and osteoblasts. Sci. Rep. 7, 15182.

Mitaka, C., Si, M.K., Tulafu, M., Yu, Q., Uchida, T., Abe, S., Kitagawa, M., Ikeda, S., Eishi, Y., Tomita, M., 2014. of atrial natriuretic peptide on inter-organ crosstalk among the kidney, lung, and heart in a rat model of renal ischemia-reperfusion injury. Intensive Care Med. Exp. 2, 28.

Molls, R.R., Rabb, H., 2004. deleterious cross-talk between failing organs. Crit. Care Med. 32, 2358−2359.

Montecalvo, A., Larregina, A.T., Shufesky, W.J., Stolz, D.B., Sullivan, M.L., Karlsson, J.M., Baty, C.J., Gibson, G.A., Erdos, G., Wang, Z., Milosevic, J., Tkacheva, O.A., Divito, S.J., Jordan, R., Lyons-Weiler, J., Watkins, S.C., Morelli, A.E., 2012. Mechanism of transfer of functional microRNAs between mouse dendritic cells via exosomes. Blood 119, 756−766.

Nitta, C.F., Orlando, R.A., 2013. Crosstalk between immune cells and adipocytes requires both paracrine factors and cell contact to modify cytokine secretion. PLoS One 8, e77306.

Nov, O., Shapiro, H., Ovadia, H., Tarnovscki, T., Dvir, I., Shemesh, E., Kovsan, J., Shelef, I., Carmi, Y., Voronov, E., Apte, R.N., Lewis, E., Haim, Y., Konrad, D., Bashan, N., Rudich, A., 2013. Interleukin-1beta regulates fat-liver crosstalk in obesity by auto-paracrine modulation of adipose tissue inflammation and expandability. PLoS One 8, e53626.

Oh, K.J., Lee, D.S., Kim, W.K., Han, B.S., Lee, S.C., Bae, K.H., 2016. Metabolic adaptation in obesity and type II diabetes: myokines, adipokines and hepatokines. Int. J. Mol. Sci. 18.

Peteranderl, C., Morales-Nebreda, L., Selvakumar, B., Lecuona, E., Vadasz, I., Morty, R.E., Schmoldt, C., Bespalowa, J., Wolff, T., Pleschka, S., Mayer, K., Gattenloehner, S., Fink, L., Lohmeyer, J., Seeger, W., Sznajder, J.I., Mutlu, G.M., Budinger, G.R., Herold, S., 2016. Macrophage-epithelial paracrine crosstalk inhibits lung edema clearance during influenza infection. J. Clin. Invest. 126, 1566−1580.

Poole, L.G., Dolin, C.E., Arteel, G.E., 2017. Organ-organ crosstalk and alcoholic liver disease. Biomolecules 7.

Robado de Lope, L., Alcibar, O.L., Amor Lopez, A., Hergueta-Redondo, M., Peinado, H., 2018. Tumour-adipose tissue crosstalk: fuelling tumour metastasis by extracellular vesicles. Philos. Trans. R. Soc. Lond. B Biol. Sci. 373.

Rodriguez, A., Becerril, S., Ezquerro, S., Mendez-Gimenez, L., Fruhbeck, G., 2017. Crosstalk between adipokines and myokines in fat browning'. Acta Physiol. 219, 362−381.

Roh, S.G., Suzuki, Y., Gotoh, T., Tatsumi, R., Katoh, K., 2016. Physiological roles of adipokines, hepatokines, and myokines in ruminants. Asian-Australas. J. Anim. Sci. 29, 1−15.

Romacho, T., Elsen, M., Rohrborn, D., Eckel, J., 2014. Adipose tissue and its role in organ crosstalk. Acta Physiol. 210, 733−753.

Shirakawa, J., De Jesus, D.F., Kulkarni, R.N., 2017. Exploring inter-organ crosstalk to uncover mechanisms that regulate beta-cell function and mass. Eur. J. Clin. Nutr. 71, 896−903.

Trayhurn, P., Drevon, C.A., Eckel, J., 2011. Secreted proteins from adipose tissue and skeletal muscle - adipokines, myokines and adipose/muscle cross-talk. Arch. Physiol. Biochem. 117, 47−56.

Valadi, H., Ekstrom, K., Bossios, A., Sjostrand, M., Lee, J.J., Lotvall, J.O., 2007. Exosome-mediated transfer of mRNAs and microRNAs is a novel mechanism of genetic exchange between cells. Nat. Cell Biol. 9, 654−659.

Virzi, G., Day, S., de Cal, M., Vescovo, G., Ronco, C., 2014. Heart-kidney crosstalk and role of humoral signaling in critical illness. Crit. Care 18, 201.

Walsh, K., 2009. Adipokines, myokines and cardiovascular disease. Circ. J. 73, 13−18.

Whitham, M., Parker, B.L., Friedrichsen, M., Hingst, J.R., Hjorth, M., Hughes, W.E., Egan, C.L., Cron, L., Watt, K.I., Kuchel, R.P., Jayasooriah, N., Estevez, E., Petzold, T., Suter, C.M., Gregorevic, P., Kiens, B., Richter, E.A., James, D.E., Wojtaszewski, J.F.P., Febbraio, M.A., 2018. Extracellular vesicles provide a means for tissue crosstalk during exercise. Cell Metabol. 27, 237−251 e4.

Wishart, D.S., Jewison, T., Guo, A.C., Wilson, M., Knox, C., Liu, Y., Djoumbou, Y., Mandal, R., Aziat, F., Dong, E., Bouatra, S., Sinelnikov, I., Arndt, D., Xia, J., Liu, P., Yallou, F., Bjorndahl, T., Perez-Pineiro, R., Eisner, R., Allen, F., Neveu, V., Greiner, R., Scalbert, A., 2013. HMDB 3.0−the human metabolome database in 2013. Nucleic Acids Res. 41, D801−D807.

Yang, H., Li, F., Xiong, X., Kong, X., Zhang, B., Yuan, X., Fan, J., Duan, Y., Geng, M., Li, L., Yin, Y., 2013. Soy isoflavones modulate adipokines and myokines to regulate lipid metabolism in adipose tissue, skeletal muscle and liver of male Huanjiang mini-pigs. Mol. Cell. Endocrinol. 365, 44−51.

Zhang, X., Ji, X., Wang, Q., Li, J.Z., 2018. New insight into inter-organ crosstalk contributing to the pathogenesis of non-alcoholic fatty liver disease (NAFLD). Protein Cell 9, 164−177.

Adipose Tissue: A Major Secretory Organ

CHAPTER OUTLINE

The Cellular Secretome and Organ Crosstalk. https://doi.org/10.1016/B978-0-12-809518-8.00002-7

For many decades, adipose tissue (AT) was considered as a passive organ being involved in the storage of lipids under conditions of excess energy intake and providing energy-rich substrates when needed by other organs. The seminal findings by Spiegelman and coworkers published between 1993 and 1996 (Hotamisligil et al., 1993, 1994a,b, 1996; Hotamisligil and Spiegelman, 1994) provided the first evidence for an active secretory function of AT, in this case, by identification of tumor necrosis factor α (TNFα) as a critical signal with auto- and endocrine functions and representing a potential link between dysregulated adipose tissue and metabolic dysfunction. Research on adipose tissue at that time and nowadays is triggered by the tremendous increase in obesity and type 2 diabetes, making it mandatory to identify novel signaling molecules and pathways that could serve as potential drug targets (for intensive discussion on secretory malfunction and its relation to metabolic diseases, see Chapter 5). The intensive research to find new therapeutic strategies to combat obesity has led to a new understanding of AT biology and the discovery of its central role in the interplay with other organs or tissues, the so-called interorgan crosstalk. (For recent reviews on adipose tissue biology, please read Huang-Doran et al., 2017; Li et al., 2017; Chen et al., 2016; Cohen and Spiegelman, 2016; Stanford and Goodyear, 2016; Townsend and Tseng, 2015; Choe et al., 2016.)

This concept of organ crosstalk was already demonstrated more than 15 years ago in studies using coculture models of human adipocytes and myocytes (Dietze et al., 2002). It is now evident that AT plays a central role in a complex and multidirectional network of autocrine/paracrine and endocrine crosstalk between organs and tissues such as the liver, skeletal muscle, vasculature, pancreas, and heart (Romacho et al., 2014). The endocrine function of AT plays a key role in orchestrating this crosstalk, with deep impact for metabolic homeostasis. Indeed, it is now thought that the downstream effects of AT-derived signaling molecules range from the central regulation of energy intake and expenditure to the control of hormone secretion from the endocrine pancreas. The reverse crosstalk back to AT is an ongoing hot topic of research, and an exciting example is presented in Chapter 3 (skeletal muscle, exercise, and the control of AT function). When talking about AT, it is mandatory to keep in mind that throughout the body different AT depots do exist, with substantially different functions and properties. At present, most of the secretome studies have been focused on white adipose tissue (WAT) (generally classified as subcutaneous or visceral depot), which is the most abundant fat tissue and plays a central role in the pathogenesis of metabolic diseases. However, there are new adipose depots arising in interorgan crosstalk. Other fat depots, such as the epicardial and perivascular AT, or even other types of AT, such as brown adipose tissue (BAT), may additionally contribute to the complexity of interorgan crosstalk (Ouwens et al., 2010; Siegel-Axel and Haring, 2016; Peschechera and Eckel, 2013).

The powerful secretory and thus endocrine function of AT is also related to the diversity of cell types being present in AT. Thus, in addition to adipocytes, preadipocytes, different types of immune cells, endothelial cells, and fibroblasts all contribute to the secretory output of AT. This comprises a large variety of molecules including metabolites, fatty acids and a variety of lipid molecules, hormones, and peptides and proteins collectively termed adipokines. Fig. 2.1 aims to summarize the secretory output of AT. As outlined in Chapter 1, the cellular secretome turns out to be extremely complex and additional mechanisms such as the release of exosomes and microvesicles do exist and most likely contribute to the crosstalk scenario. In this chapter, as overall in this book, we focus on the protein factors released within and from different AT depots, considering this setting as a paradigm of crosstalk regulation. However, we do not want to distract from the importance of other molecules in addition to adipokines, but their functional impact is beyond the scope of this book.

Adipokines comprise, among others, classical cytokines and chemokines, vasoactive and coagulation factors, regulators of lipoprotein metabolism, and proteins more specifically secreted by the adipocytes, such as leptin or adiponectin (Romacho et al., 2014; Wronkowitz et al., 2014; Fasshauer and Bluher, 2015). Since the discovery of leptin, the number of adipokines has notably increased in the last years with new molecules such as visfatin, apelin, omentin, or dipeptidyl peptidase 4 (DPP4) among others (Lamers et al., 2011; Rohrborn et al., 2015). Recently, our group demonstrated the complexity of the human adipokinome consisting of hundreds of different factors (Lehr et al., 2012). In this study, it was demonstrated that human adipocytes release more than 600 putative adipokines, making it likely that the secretory output of AT exceeds 1000 adipocytokines, taking into account the presence of immune and other cell types in this tissue. Many of these adipokines can act locally within the AT, and this autocrine/paracrine crosstalk plays a significant role for the immunoregulatory function and plasticity of AT (Sell et al., 2012). From the host of adipokines produced within AT, a certain part can also reach distant organs through the systemic circulation, where they can exert a wide range of biological actions, including the regulation of food intake and body weight, insulin sensitivity, inflammation, coagulation, or vascular function (Romacho et al., 2014; Stern et al., 2016). Thus, adipokines represent key signaling and mediator molecules for AT

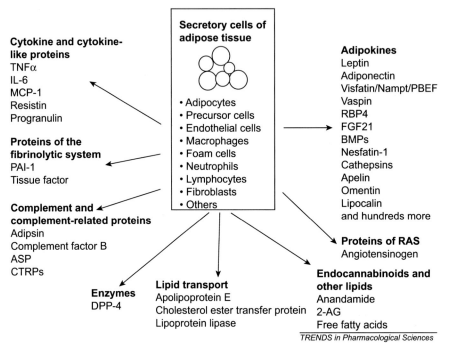

Secretory cells of adipose tissue

- Adipocytes
- Precursor cells
- Endothelial cells
- Macrophages
- Foam cells
- Neutrophils
- Lymphocytes
- Fibroblasts
- Others

Cytokine and cytokine-like proteins
TNFα
IL-6
MCP-1
Resistin
Progranulin

Proteins of the fibrinolytic system
PAI-1
Tissue factor

Complement and complement-related proteins
Adipsin
Complement factor B
ASP
CTRPs

Enzymes
DPP-4

Lipid transport
Apolipoprotein E
Cholesterol ester transfer protein
Lipoprotein lipase

Adipokines
Leptin
Adiponectin
Visfatin/Nampt/PBEF
Vaspin
RBP4
FGF21
BMPs
Nesfatin-1
Cathepsins
Apelin
Omentin
Lipocalin
and hundreds more

Proteins of RAS
Angiotensinogen

Endocannabinoids and other lipids
Anandamide
2-AG
Free fatty acids

TRENDS in Pharmacological Sciences

FIGURE 2.1

Secretory output from adipose tissue. *BMPs*, bone morphogenetic proteins; *DPP4*, dipeptidyl peptidase 4; *FGF21*, fibroblast growth factor 21; *IL-6*, interleukin-6; *MCP-1*, monocyte chemoattractant protein 1; *Nampt*, nicotinamide phosphoribosyltransferase; *PAI-1*, plasminogen activator inhibitor-1; *PBEF*, pre-B cell colony-enhancing factor; *RBP4*, retinol-binding protein 4; *TNFα*, tumor necrosis factor α.

Reproduced with permission from Fasshauer, M., Bluher, M., 2015. Adipokines in health and disease, Trends Pharmacol. Sci. 36, 461–470.

to establish a complex network of feedback loops with other distant target organs or tissues. This crosstalk scenario fueled by adipokines plays an important role for metabolic homeostasis under healthy conditions, whereas the secretory malfunction of enlarged AT involving a prominent release of proinflammatory adipocytokines is detrimental and a key driver in the pathogenesis of many metabolic disorders. This topic will be discussed in detail in Chapter 5.

Because adipokines can be considered as paradigm-secreted polypeptide molecules, we will start in this chapter with a deeper look on the different pathways of adipokine release, including the classical ER (endoplasmic reticulum)/Golgi route, nonclassical pathways, and the so-called shedding of proteins from the cell surface (see Fig. 2.2). This chapter also provides a detailed introduction into the autocrine/paracrine and endocrine crosstalk pathways with emphasis on selected adipokines

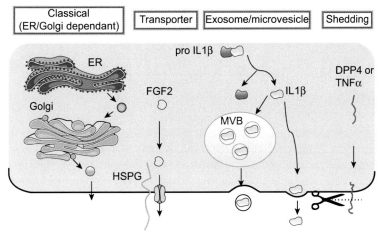

FIGURE 2.2

Routes of protein release. *DPP4*, dipeptidyl peptidase 4; *ER*, endoplasmic reticulum; *FGF2*, fibroblast growth factor 2; *IL1β*, interleukin-1β; *MVB*, multivesicular bodies; *TNFα*, tumor necrosis factor α.

Reproduced with permission from Romacho, T., Elsen, M., Rohrborn, D., Eckel, J., 2014. Adipose tissue and its role in organ crosstalk. Acta Physiol. (Oxf) 210, 733-753.

and the fat—muscle, fat—liver, fat—pancreas axis. Finally, we will focus on the role of specific fat depots such as brown fat, perivascular fat, and epicardial fat, different tissues that have gained considerable interest and contributed to a remarkable extension of the crosstalk concept.

MECHANISMS OF ADIPOKINE RELEASE

Although the number of identified adipokines is continuously increasing, our knowledge regarding the precise mechanisms of adipokine release remains rather limited, and more detailed data are only available for a small percentage of the large number of adipokines. In general, proteins that contain an N-terminal signal peptide are exported via a so-called classical or ER/Golgi-dependent secretory pathway. In addition to this classical export route, there is an increasing number of adipokines that are secreted through alternative routes. These proteins are characterized by a lack of the signal peptide sequence and by insensitivity to brefeldin A or monensin treatment (Nickel, 2003). Evidence exists that adipokines can be released by constitutive secretion, regulated, or both. In the following sections, we take a detailed look on the different routes of protein export from AT. As examples of adipokines secreted by the classical mechanism, we have chosen two prototypes of adipokines, namely adiponectin and leptin. The multiple functions of these adipokines are described in detail in the later sections of this chapter. Adipokines released through nonclassical secretory mechanisms are introduced with specific examples of transporter-mediated export such as that of FGF2 (fibroblast growth factor 2), shedding such as TNFα and DPP4, and multiple secretion routes such as interleukin (IL)-1β.

CLASSICAL OR ENDOPASMIC RETICULUM/GOLGI-DEPENDENT PATHWAYS

The classical ER/Golgi secretion mechanism comprises proteins that contain a signal peptide to lead their translocation to the ER and to mediate further secretion (Palade, 1975). In brief, mRNA of the respective protein is translated at the ribosome of the rough ER, and subsequently, the protein is translocated into this compartment. Inside the ER, the correct folding and assembly of the proteins is assured. Via COPII-coated vesicles, proteins exit the ER and enter the Golgi complex, where they are further processed and sorted for transport to their destination. Subsequently, vesicles fuse with the plasma membrane and release the proteins into the extracellular space. For additional reading on this topic, we recommend Gomez-Navarro and Miller (2016) and Zacharogianni and Rabouille (2013). Two of the most studied adipokines, leptin, and adiponectin are secreted from adipocytes using this pathway.

ADIPONECTIN

Adiponectin is one of the most abundant adipokines and was first described by Scherer et al. (1995). This adipokine exhibits a unique action profile and exerts multiple beneficial effects on a number of target tissues (Stern et al., 2016). These effects of adiponectin will be described in the sections to follow. There are three secreted oligomeric isoforms of adiponectin: trimeric, hexameric, and high-molecular weight (HMW). Adiponectin multimerisation and secretion is initially controlled via thiol-mediated retention in the transit through the ER. This retention is mediated via direct interactions with chaperones such as ERp44, whereas the oxidoreductase Ero1-Lα interacts with ERp44, promoting adiponectin release. Expression of both proteins is tightly regulated by the metabolic status of the cell, providing a feedback loop for adiponectin release. Thus, proliferator—activated receptor γ (PPARγ) activation enhances Ero1-Lα gene expression leading to increased HMW-adiponectin secretion in 3T3-L1 adipocytes (Wang et al., 2008). When moving through the Golgi and *trans*-Golgi network, a pool of adiponectin molecules is packaged into GGA1 (Golgi localizing γ-adaptin ear homology domain ADP ribosylating factor[ARF] binding)-coated vesicles and delivered to endosomes (Xie et al., 2006).

It is important to note that insulin stimulates adiponectin release in rat adipocytes, and this may be mediated by activation of the endosomal membrane recycling with the participation of Rab5 and Rab11. These small GTPases are markers for the early/sorting and recycling compartments, respectively. Interestingly, it was recently demonstrated that the novel protein FIP1 (family of interacting proteins 1), an effector of Rab11, negatively regulates adiponectin trafficking and secretion (Carson et al., 2013). FIP1 expression is downregulated during adipogenesis and by thiazolidinediones, whereas TNFα upregulates FIP1 expression in 3T3L1 adipocytes. These data fit very well to the observed regulation of adiponectin secretion by different stimuli and provide a molecular mechanism underlying these effects.

LEPTIN

Leptin, the product of the *ob* gene, was discovered more than 20 years ago and is the first adipokine to be identified (Zhang et al., 1994). Leptin exerts a prominent action on the hypothalamus with a key role in regulating food intake, energy expenditure, and neuroendocrine function among many others (Flak and Myers, 2016). Localization analysis by velocity and density gradient ultracentrifugation identified leptin in the high-density microsomes, which are positive for ER markers (Cammisotto et al., 2005). As a classically transported adipokine, leptin secretion is abolished by brefeldin A treatment. Mechanistically, it was proposed that the serine/threonine protein kinase D1 might mediate the fission at the *trans*-Golgi network of leptin transport vesicles to the plasma membrane (Xie et al., 2008). Regulated secretion of leptin in adipocytes was found to be highly induced by a mixture of glucose, insulin, and pyruvate. The concentrations used mimicked hyperglycemia and hyperinsulinemia, and leptin secretion was stimulated through vesicular exocytosis. Both the constitutive secretion and the stimulated secretion of leptin were found to be Ca^{2+}-dependent. Moreover, forskolin, an activator of adenylate cyclase, abolished the facilitating effects of the glucose, insulin, and pyruvate mixture, suggesting that cAMP/PKA may act downstream of these molecules. Insulin is thought to increase the translation and translocation of leptin through the PI3K/mTOR signaling pathway. Accordingly, insulin-stimulated leptin secretion is inhibited by β-*agonists* via adenylate cyclase activation.

NONCLASSICAL PROTEIN SECRETION

Proteins lacking an N-terminal signal peptide leading to ER/Golgi transport can also be secreted by alternative, so-called nonclassical pathways. Several well-known adipokines, such as TNFα, visfatin, Pref1, or DPP4, are released through these mechanisms from AT. To date, there are at least three distinct nonclassical release pathways known, namely transporter-mediated export, microvesicle/exosome release, and the selective posttranslational hydrolysis from the cell surface, which is also known as "shedding". FGF2, DPP4, and IL-1β are well-known examples for these different mechanisms, and the following sections summarize our knowledge on the secretion of these adipokines.

TRANSPORTER-MEDIATED EXPORT OF FGF2

The adipokine FGF2 (, also known as basal FGF) is mainly released by preadipocytes and plays a role in adipogenesis (Hutley et al., 2011; Kakudo et al., 2007; Xiao et al., 2010). As pointed out before, FGF2 lacks a transient signal sequence, and its release is insensitive to brefeldin A treatment. Methylamine, which inhibits exocytosis, showed a decrease in FGF2 release, indicating externalization via a mechanism of exocytosis independent of the ER/Golgi complex (Mignatti et al., 1992). Experiments using ouabain, an inhibitor of the plasma membrane Na-K-ATPase, support an export mechanism through this transporter (Dahl et al., 2000; Florkiewicz et al., 1995). It was also shown that FGF2 is recruited to the inner leaflet of the plasma membrane before it is translocated to the extracellular space (Schafer et al., 2004).

SHEDDING OF TNFα AND DPP4

Accumulating evidence now strongly suggests that TNFα is one important mediator, which triggers obesity-related insulin resistance. During obesity, the expression of TNFα is upregulated in AT, muscle, and specifically macrophages. TNFα might act mainly in an autocrine/paracrine fashion at its target sites in obesity (for a detailed consideration on autocrine/paracrine crosstalk, see sections below). By proteolytic cleavage at Ala76-Val77 in the extracellular part of the transmembrane protein, the soluble form of TNFα is released. This release step is affected by hydroxamic acid–based inhibitors, which are broad-spectrum inhibitors for matrix metalloproteases (MMP). This indicates that the release of TNFα is mediated by at least one or most likely by several MMPs. Black and colleagues cloned and purified TACE (TNFα—*converting enzyme;* also known as Adam17) and identified TACE as the major shedding enzyme of TNFα (Black et al., 1997).

DPP4 is a ubiquitously expressed type II transmembrane protease, which is able to cleave and inactivate N-terminal dipeptides from a variety of substrates, including growth factors and hormones (such as the incretin hormones), neuropeptides, and chemokines (Rohrborn et al., 2015). Importantly, not only DPP4 is present on the cell surface, but there also exists a soluble form of this enzyme, which is found in the circulation. Work conducted at the German Diabetes Center in Düsseldorf identified DPP4 as a novel adipokine by proteomic profiling of the adipocyte secretome (Lamers et al., 2011). Strikingly, augmented serum levels of DPP4 were found in obese patients and correlate with the size of adipocytes and risk factors for the metabolic syndrome. Thus, it was concluded that DPP4 may represent a novel marker for visceral obesity and a potential link to the metabolic syndrome (Sell et al., 2013). Like most type II transmembrane proteins, DPP4 contains a predicted signal sequence located at the transmembrane domain (revealed by bioinformatic analysis with the SignalP webtool). It was observed, however, that the release of DPP4 is insensitive to brefeldin A treatment (Rohrborn et al., 2014). It is most likely that DPP4 is released by ectodomain shedding, which means that the extracellular part of this type II transmembrane protein is selectively hydrolyzed from the cell surface (Romacho et al., 2014). Currently available data suggest that MMP9 could represent an important player in the shedding process of DPP4 (Rohrborn et al., 2014).

MULTIPLE EXPORT ROUTES OF IL-1β

It is known for longtime that IL-1 lacks a classical signal peptide (Dinarello, 1987). Both isoforms of IL-1 are secreted in an unconventional way, but much more is known about IL-1β release. The mechanism of production and maturation of IL-1β has been studied extensively; however, the mechanism by which IL-1β is released into the extracellular space is still controversially discussed. At least four possible active export routes have been reported (Eder, 2009). Rubartelli and coworkers suggested three of these routes in several studies on IL-1β in human macrophages, including different kinds of vesicles, which might be lysosomes, shed plasma membrane vesicles, or exosomes from multivesicular bodies (Carta et al., 2013). It was also shown that the ABC transporter inhibitor glyburide selectively inhibits IL-1β release in murine macrophages, making it likely that IL-1β is exported via an active, transporter-mediated process (Brough and Rothwell, 2007). These multiple export routes of IL-1β emphasize the complexity of adipocytokine secretion and release, making it possible that multiple regulatory control loops exist at this level, an important issue for future research and a potential yet unexplored target to control adipokine secretion.

Auto- and Paracrine Crosstalk Within Adipose Tissue and the Regulation of Adipogenesis

AT is characterized by an amazing and unique plasticity reflected by the percentage of body weight representing less than 5% up to over 60% in some individuals. Depending on the depot involved, this plasticity makes it possible to store large amounts of lipids without perturbation of the metabolic performance, leading to the so-called setting of healthy obesity (Bluher, 2010, 2012, 2014). Expansion of AT involves increased lipid loading of mature adipocytes that are able to increase their regular size severalfold (AT hypertrophy) (Bays et al., 2008).

Adipocyte size itself is linked with metabolic dysfunction (Luo and Liu, 2016), and enlarged adipocytes may contribute to the development of metabolic dysfunction by a decreased lipid-buffering capacity, leading to ectopic fat accumulation. In addition, there is a shift toward a more proinflammatory secretion profile with increased adipocyte size (Skurk et al., 2007). The dysregulated secretome of enlarged adipocytes and the impact of this secretory malfunction will be addressed later in Chapter 5, in which, we will also discuss the modulation of AT inflammation by adipokines and the immunoregulatory function of AT in the context of obesity and metabolic dysfunction.

A second arm involved in AT plasticity and expansion relates to an increased differentiation of adipocyte precursor cells resulting in an increased adipocyte number (hyperplastic growth), a process referred to as adipogenesis. This process involves changes in morphology, insulin sensitivity, and the secretome of these cells (Ali et al., 2013; Sarjeant and Stephens, 2012; Tang and Lane, 2012). It is important to note the significant progress obtained in the methodologies, genetic tools, and characterization of adipocyte precursor cells and their impact for adipogenesis,

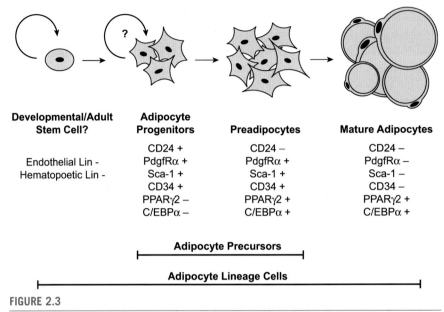

Developmental/Adult Stem Cell?	Adipocyte Progenitors	Preadipocytes	Mature Adipocytes
Endothelial Lin - Hematopoetic Lin -	CD24 + PdgfRα + Sca-1 + CD34 + PPARγ2 − C/EBPα −	CD24 − PdgfRα + Sca-1 + CD34 + PPARγ2 + C/EBPα +	CD24 − PdgfRα − Sca-1 − CD34 − PPARγ2 + C/EBPα +

Adipocyte Precursors

Adipocyte Lineage Cells

FIGURE 2.3

A model of in vivo adipogenesis.

With permission from Berry, R., Jeffery, E., Rodeheffer, M.S., 2014. Weighing in on adipocyte precursors. Cell Metab. 19, 8–20.

although it is beyond the scope of this chapter (and this book) (Berry et al., 2014). This is of specific importance, as there is a constant turnover in the number of adipocytes during life with ~10% of all adipocytes being renewed each year (Spalding et al., 2008). Thus, de novo adipogenesis is crucial to maintain adipocyte number and to provide new adipocytes in case of excess energy supply. A model of in vivo adipogenesis is depicted in Fig. 2.3 and is largely based on the analysis of adipogenic cell populations in mice. As suggested by Rodeheffer and coworkers (Berry and Rodeheffer, 2013), the CD24[+] population contains early adipocyte progenitors (mostly derived from mesenchymal stem cells [MSCs]) with high adipogenic capacity within the AT microenvironment. The CD24[+] preadipocytes are committed to the adipocyte lineage and differentiate to mature adipocytes.

The differentiation of preadipocytes to adipocytes involves a complex network of intraorgan crosstalk and transcription factors that allow the expression of key proteins leading to mature adipocyte formation (see Fig. 2.4). Differentiation into adipocytes requires a temporally regulated transcriptional cascade. Key transcription factors in adipogenesis are the nuclear receptor peroxisome PPARγ and the CCAAT/ enhancer-binding proteins (C/EBPs) (Lefterova and Lazar, 2009). The transcriptional cascade can be influenced by several pro- and antiadipogenic secretory products of the different cell types present in AT (such as immune cells, preadipocytes, adipocytes).

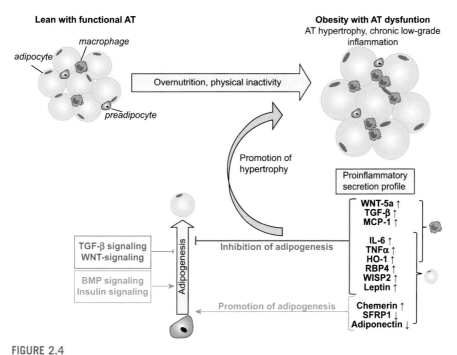

FIGURE 2.4

Regulation of adipogenesis by adipocytokines. *AT*, adipose tissue; *BMP*, bone morphogenetic protein; *IL1β*, interleukin-1β; *RBP4*, retinol-binding protein 4; *SFRP1*, secreted frizzled-related protein 1; *TGFβ*, transforming growth factor β; *TNFα*, tumor necrosis factor α; *WISP2*, WNT1-inducible signaling pathway protein 2.

With permission from Romacho, T., Elsen, M., Rohrborn, D., Eckel, J., 2014. Adipose tissue and its role in organ crosstalk. Acta Physiol. (Oxf) 210, 733–753.

Bone morphogenetic proteins (BMPs) promote adipogenesis (Schulz and Tseng, 2009), whereas TGF-β (transforming growth factor β) inhibits adipogenesis (Zamani and Brown, 2011). Another inhibitory pathway in the regulation of adipocyte differentiation is the Wnt signaling pathway, which is activated by the Wnt family of secreted glycoproteins (see below for detailed discussion of this pathway). Cytokines released from macrophages and hypertrophic adipocytes, such as TNFα and IL-6, have been shown to inhibit adipocyte differentiation. It is well known that the classical adipokines leptin and adiponectin modulate adipogenesis in an inverse fashion, with leptin being inhibitory and adiponectin being a stimulator of fat cell differentiation. Most of the adipokines that inhibit de novo adipogenesis are upregulated in the obese state, whereas promoters of adipogenesis such as adiponectin are downregulated (Fig. 2.4). In this section, we start with a look on leptin and adiponectin regarding their autocrine/paracrine functions with emphasis on the regulation of adipogenesis. We will then turn to novel AT-derived factors regulating adipogenesis in an autocrine/paracrine manner. More specifically we will address the autocrine/paracrine effects of several adipokines, which are upregulated in obesity and exert a negative effect on adipogenesis such as retinol-binding protein 4 (RBP4), WNT1-inducible signaling pathway protein 2 (WISP2), and the recently identified adipokine HO-1. Furthermore, we summarize here the main effects of novel adipokines that may positively regulate adipogenesis such as secreted frizzled-related proteins (SFRPs) and chemerin.

ADIPONECTIN AND LEPTIN

Both leptin and adiponectin exhibit a prominent endocrine function (see subsequent sections), but both adipokines do also act in an autocrine/paracrine fashion within AT and exert an important regulatory effect on AT lipid metabolism.

Leptin is a 16 kDa polypeptide encoded by the *ob* gene and originally described in 1994 by Friedman and coworkers (Zhang et al., 1994). Although initially described as a satiety hormone, leptin can be considered as the paradigm adipokine, and the identification of leptin was definitely the starting point for a completely new view on AT function. Leptin is secreted by adipocytes, signals via the leptin receptor, and exerts multiple functions on appetite, food intake, fertility, and fetal growth (Mantzoros et al., 2011). Leptin receptors are also present on adipocytes, and hence an autocrine action of leptin on adipose cells could be demonstrated in vitro including increased lipolysis and increased fatty acid oxidation (William et al., 2002). However, more recent data indicate that in addition to this autocrine effect, leptin-induced lipolysis is mediated by sympathetic neurons innervating adipocytes (Zeng et al., 2015). Another autocrine function of leptin relates to the inhibition of adipogenesis (Rhee et al., 2008). As reported by these authors, acute and prolonged treatment of preadipocytes with leptin inhibited adipocyte differentiation along with decreased expression of PPARγ and aP2, two key transcription factors mediating adipogenesis. These actions appear to be mediated by MAP kinase and JAK/STAT signaling pathways (Rhee et al., 2008).

Adiponectin was discovered by Scherer and coworkers (Scherer et al., 1995), and this 30 kDa polypeptide is one of the most highly expressed proteins almost exclusively produced by adipocytes. Adiponectin has gained considerable interest owing to its multiple beneficial effects including insulin-sensitizing, antiinflammatory, and antiapoptotic effects (Stern et al., 2016). The circulating level of adiponectin is extremely high (about 0.01% of all plasma proteins), and adiponectin exists in three different species owing to posttranslational modifications: a low-molecular-weight trimer, found in the circulation as 180 kDa hexamers, and a HMW multimer of more than 300 kDa (Turer and Scherer, 2012). Adiponectin signals via two different receptors, AdipoR1 and AdipoR2, and these receptors define a certain tissue specificity of adiponectin. The multiple and highly important complex endocrine functions of adiponectin will be addressed in the subsequent sections including crosstalk from fat to the muscle, liver, pancreas, and vasculature. Different studies in animal models suggested an autocrine function of adiponectin owing to the observation of a healthy AT expansion and reduced ectopic lipid storage (Xu et al., 2003).

The first evidence for an autocrine action of adiponectin on human fat cells was reported in 2005 by the Eckel laboratory (Dietze-Schroeder et al., 2005), showing that adiponectin acts as a master regulator of adipocyte secretory function. In the same year, the Garvey laboratory published a paper showing another autocrine function of adiponectin: the promotion of preadipocyte proliferation and differentiation, involving enhanced expression of genes responsible for adipogenesis (Fu et al., 2005). These results were obtained by overexpression of adiponectin in 3T3-L1 fibroblasts. Subsequently, the proadipogenic action of adiponectin was confirmed using human preadipocytes (do Carmo Avides et al., 2008).

RETINOL-BINDING PROTEIN 4

RBP4 belongs to the lipocalin family of transport proteins. The primary site of RBP4 synthesis is the liver, but AT is also an important source for this protein (Tsutsumi et al., 1992). It has been demonstrated that mature adipocytes represent the main source of RBP4 within AT (Cheng et al., 2013). RBP4 was identified as an adipokine in 2005, and augmented serum RBP4 levels were observed in several obese and diabetic mouse models (Yang et al., 2005). However, so far the precise role of RBP4 as a molecular pathway to insulin resistance has remained controversial. About 90% of circulating RBP4 is bound to retinol (holo-RBP4), and 10% is present in the unbound form (apo-RBP4). As retinol derivates, retinoids activate the nuclear receptor retinoid acid receptor (RAR) and retinoid X receptor (RXR), two factors involved in the regulation of adipogenesis (Ziouzenkova and Plutzky, 2008). RBP4 overexpression in preadipocytes has been shown to decrease adipogenesis, mediated by impaired AKT and mTOR phosphorylation (Cheng et al., 2013). Differentiation of 3T3-L1 preadipocytes is blocked by retinol-bound holo-RBP4, involving increased RARα activation. This effect is dependent on retinol, as treatment of 3T3-L1 preadipocytes with the retinol-free apo-RBP4 increased retinol efflux and enhanced differentiation (Muenzner et al., 2013).

An additional level of complexity results from the regulation of RBP4 expression in AT. Several adipocyte- and macrophage-derived factors are able to control RBP4 expression in an autocrine/paracrine manner. Thus, RBP4 protein expression is downregulated in primary human adipocytes by TNFα (Sell and Eckel, 2007) and IL-1β (Kotnik et al., 2013). Interestingly, leptin increases RPB4 protein level in human AT explants. The insulin-sensitizing PPARγ agonists troglitazone and pioglitazone have been shown to increase RBP4 expression in human adipocytes (Sell and Eckel, 2007). This is in conflict with the postulated role of RBP4 in the induction of insulin resistance and requires further investigation. RBP4 might be dysregulated in states of AT inflammation because a positive association between expression of RBP4 and the macrophage marker CD68 was found in subcutaneous human AT (Yao-Borengasser et al., 2007). RBP4 also directly acts on macrophages and impairs macrophage secretory function. Macrophages challenged with apo-RBP4 or holo-RBP4 were shown to express increased levels of MCP1 (monocyte chemoattractant protein 1), IL-6, and TNFα, which were independent of retinol and the RBP4 receptor STRA6 (Norseen et al., 2012). In conclusion, the adipokine RBP4 may contribute to the development of metabolic dysfunction by impairing adipocyte differentiation and increasing secretion of proinflammatory cytokines by macrophages. The RBP4-mediated induction of TNFα may further inhibit adipocyte differentiation. However, as pointed out before, it remains to be explained why RBP4 expression in adipocytes is decreased by TNFα and increased by PPARγ agonists. Moreover, some of the described effects of RBP4 are dependent on retinol and some are not, complicating the role of RBP4 in AT function.

HEME OXYGENASE-1

Heme oxygenases (HOs) cleave heme to biliverdin, iron, and carbon monoxide (CO) and can be found as two isozymes. While HO-2 is constitutively expressed in a number of different tissues, HO-1 is a stress-responsive isoform inducible by oxidative stress, cytokines, UV light, and other factors. Increased HO-1 activity protects against oxidative stress, as present in diabetes and atherosclerosis (Abraham and Kappas, 2008). Using a proteomics approach, it was shown that HO-1 is secreted from primary human adipocytes and HO-1 protein levels increase during adipocyte differentiation. HO-1 protein abundance is substantially higher in adipocytes compared with macrophages; therefore, this adipokine is rather an adipocyte-derived factor in AT (Lehr et al., 2012). The protein level of HO-1 is increased in subcutaneous and visceral fat from obese patients compared with that from matched lean controls. It was also found that the serum HO-1 level is upregulated in human obesity (Lehr et al., 2012) and type 2 diabetic subjects (Bao et al., 2010). More recent studies demonstrated positive effects of HO-1 on AT function. Induction of HO-1 by cobalt protoporphyrin IX (CoPP) injection prevents AT inflammation in obese animals (Kim et al., 2008) and lowers weight gain and body fat content, indicating a role for HO-1 in the regulation of adipogenesis. In line with this, adipogenic differentiation is increased in MSCs from HO-2− deficient mice, which display lower HO-1 protein expression. Similar effects on adipogenesis and adipokine secretion were observed in human MSCs. When these cells were treated with the HO-1 inducer CoPP, an increased number of small lipid droplets were found, whereas the number of large lipid droplets was reduced. The inhibitory effect of HO-1 on adipogenesis could be mediated by increased expression of the WNT-signaling components Wnt10b and β-catenine (Vanella et al., 2013). Despite clear evidence for inhibition of adipogenesis, HO-1 seems to improve the adipokine secretion profile and to prevent obesity induced by high-fat diet in animal models (Kim et al., 2008). However, a recent study showed that selective overexpression of HO-1 in AT could not prevent diet-induced obesity, insulin resistance, and the expression of proinflammatory cytokines (Huang et al., 2013). Thus, HO-1 appears to be a player in the early induction of adipogenesis, as HO-1 induction in preadipocytes decreases adipogenesis (Vanella et al., 2013).

WNT1-INDUCIBLE SIGNALING PATHWAY PROTEIN 2

WISP2, also known as CCN5, is a member of the CCN (connective tissue growth factor/cysteine-rich 61/nephroblastoma overexpressed) family and is induced by Wnt/β-catenine signaling (Jackson et al., 2005). Extracellular WNT ligands are known to activate canonical WNT/β-catenine signaling, leading to β-catenine stabilization and interaction with the nuclear transcription factors TCF/LEF. Inhibition of WNT/β-catenine signaling is a mandatory step to induce adipogenic differentiation. Thus, harmine-mediated induction of 3T3-L1 preadipocyte differentiation is accompanied by inhibition of WISP2 and other components of the Wnt-signaling pathway, suggesting a role for WISP2 in the regulation of adipogenesis (Waki et al., 2007). We recently analyzed the fat cell secretome, considering expression of adipokines during adipogenesis and their regulation in obesity, and identified WISP2 as a top candidate upregulated in obesity (Dahlman et al., 2012). Confirming these observations, WISP2 was recently validated and characterized as novel adipokine (Hammarstedt et al., 2013). Importantly, expression of WISP2 mRNA in subcutaneous AT from human subjects correlates positively with adipocyte size and waist circumference. It was also shown that the knockdown of WISP2 in 3T3-L1 cells is able to induce spontaneous differentiation of these cells, whereas treatment with recombinant WISP2 suppresses adipogenesis (Hammarstedt et al., 2013). Thus, WISP2 is a potent new regulator of adipocyte differentiation (see Fig. 2.4). It is important to note that the positive regulator of white adipocyte differentiation BMP4 is able to counteract the inhibitory effect of WISP2 on adipogenesis. Future studies need to determine whether WISP2 is secreted by other cell types present in AT and how this process is regulated in states of AT inflammation.

SECRETED FRIZZLED-RELATED PROTEINS

Members of the family of SFRPs are extracellular regulators of Wnt signaling, a process involving binding to and changing the activity of Wnt ligands (Park et al., 2008). The SFRP family comprises five members in mammals (SFRP1-5). SFRP1 expression increases during adipogenesis, and it is primarily found in mature adipocytes, indicating a role in the regulation of adipogenesis. In fact, overexpression of SFRP1 in 3T3-L1 preadipocytes is able to inhibit the canonical WNT-signaling pathway and to stimulate adipogenesis (Lagathu et al., 2010). SFRP1 is involved in the regulation of adipocyte number during development of AT hypertrophy and dysfunction. Thus, SFRP1 expression initially rises in mice under high-fat diet to compensate the increased demand for fat storage, but it decreases on development of AT dysfunction and metabolic complications (Lagathu et al., 2010). In line with this, SFRP1 expression in subcutaneous AT is increased in mild obese humans and falls to levels equal to those in lean subjects in morbidly obese humans.

In addition to SFRP1, SFRP2, and SFRP4, there are two members of the SFRP family that are also able to promote adipogenesis (Park et al., 2008). However, the recent study by Ehrlund et al. (2013) could not observe any effect of SFRP2 and SFRP4 on adipogenesis in human adipose-derived stem cells (hADSCs). The role of SFRP5 as an adipokine and its impact for adipogenesis has remained very contro-versial. SFRP5 expression has been reported to be decreased in several obese mouse models and in visceral AT of obese individuals (Ouchi et al., 2010) and was described as an antiinflammatory adipokine-modulating metabolic dysfunction. $Sfrp5^{-/-}$ mice under high-fat diet develop insulin resistance and AT inflammation and have enlarged adipocytes, providing strong evidence that SFRP5 promotes adipogenesis and has antiinflammatory properties (Ouchi et al., 2010). In contrast, SFRP5 has been shown to be expressed in human AT explants but was found not to be secreted time-dependently from these explants and therefore was not consid-ered to be a real adipokine (Ehrlund et al., 2013). In summary, most studies observed a positive effect of SFRPs on adipogenesis, most likely reflecting their function as negative regulators of WNT signaling. SFRPs can be considered as very promising targets, and their regulation and function in AT has to be further investigated.

CHEMERIN

Chemerin (RARRES2 or TIG) is a chemoattractant protein and binds to chemokine-like receptor 1 (CMKLR1), playing a role in innate and adaptive immunity (Bondue et al., 2011). The WAT, liver, and placenta express high levels of chemerin mRNA, and this polypeptide has recently been identified as a novel adipokine secreted by 3T3-L1 adipocytes (Goralski et al., 2007). This adipokine is mainly derived from mature adipocytes, as expression of chemerin mRNA is strongly upregulated during differentiation of preadipocytes (Sell et al., 2009). As observed for many other adipo-kines, chemerin is dysregulated in obesity and has been proposed as a potential link between obesity and T2DM. Thus, expression of chemerin and its receptor CMKLR1 is increased in AT from obese rats compared with that of lean controls. In obese hu-man subjects, an augmented serum level of chemerin was reported, and secretion of chemerin was higher from obese human AT explants (Sell et al., 2009). Importantly, data suggest that chemerin is another factor regulating adipogenesis in an autocrine/paracrine manner. Knockdown of chemerin in 3T3-L1 preadipocytes (Goralski et al., 2007) abrogates adipogenesis, suggesting that chemerin is necessary for adipocyte dif-ferentiation. Chemerin mainly affects the mitotic clonal expansion phase in the initial phase of differentiation and has no effect on adipogenesis when knocked down in later phases of differentiation (Goralski et al., 2007). An inverse regulation of chemerin expression is observed when comparing preadipocytes with mature adipocytes. Thus, PPARγ agonists enhance chemerin expression in preadipocytes, whereas treat-ment of mature fat cells with PPARγ agonists decreases chemerin mRNA expression and release (Sell et al., 2009). Furthermore, chemerin release from human adipocytes is increased by the proinflammatory cytokine TNFα. Accumulation of lipids during differentiation increases chemerin protein levels (Bauer et al., 2011). In summary, current evidence indicates that chemerin is a positive regulator of adipogenesis and is upregulated in obesity. Enhanced expression of chemerin in adipocytes with enlarged lipid droplets may be acting as a feedback loop to promote adipogenesis and thereby provide new adipocytes.

Adipose Tissue and Its Endocrine Function

The identification of the paradigm adipokines leptin and adiponectin was coupled to the recognition that AT represents a key endocrine organ that exerts major metabolic and immunomodulatory functions. Since then a number of new fat-derived hormones have been described. These molecules enter the circulation, bind to their appropriate receptors present on target cells, and, in that way, regulate metabolic pathways and transcriptional networks in multiple organs including the skeletal muscle, liver, vasculature, heart, brain, pancreatic islets, skin, immune cells, and many others. Despite some earlier works on the endocrine function of steroid hormones released by AT, the identification of leptin and the discovery of its hypothalamic function leading to reduced appetite must be considered as a paradigm of AT-mediated endocrine crosstalk (Mantzoros et al., 2011). This concept has been substantially extended over the years, and today a highly complex scenario exists involving not only multiple organs addressed by AT secretory products, but also an inverse communication and backward loops, finally orchestrating the fine-tuning of metabolic control. However, some caution is needed regarding in vitro data on the crosstalk between adipocytes or AT. In a substantial number of studies, conditioned media (CM) are generated from adipocytes or AT samples and are then applied to a large array of other cells ranging from macrophage cell lines to primary human cells obtained from different organs. Although of some interest, it remains questionable whether this kind of approach would really mimic the in vivo endocrine crosstalk of adipokines. More specifically, in many studies, it remains unclear whether the factors present in adipocyte-CM are really present in the circulation and hence can really exert endocrine functions. For a new adipokine, in vivo experiments, specifically in humans, need to be conducted to provide evidence for release to the circulation. As an example, the paper by Sell and coworkers (Sell et al., 2013) demonstrating depot-specific release of the novel adipokine DPP4 should be considered.

In the following sections, we wish to focus on three different highly important crosstalk pathways emanating from AT, specifically involving not only the visceral but also the subcutaneous fat depot. We start with skeletal muscle, one of the largest organs in the body with deep impact for glucose homeostasis and insulin action. The negative crosstalk from fat to muscle in obesity gained huge interest at the beginning of crosstalk research in this field, and it is now considered as a key driver in the pathogenesis of insulin resistance. The fat–liver crosstalk also affects insulin sensitivity, but potentially fatty acid flux to the liver and lipotoxic mediators play an additional role in this organ communication. We finally look on the axis from AT to the pancreatic β-cell, a very exciting and more recently extensively analyzed crosstalk scenario.

THE FAT–MUSCLE CROSSTALK

Skeletal muscle plays a key role in insulin sensitivity as a main target for insulin-stimulated glucose disposal. This involves activation of a complex downstream signaling pathway triggered by autoactivation of the insulin receptor, a subsequent phosphorylation cascade leading to activation of Rab-GTPases, which finally mediates the translocation of the glucose transporter GLUT4 to the plasma membrane (Deshmukh, 2016). Insulin signaling involves a highly complex machinery. Intracellular signaling crosstalk between the insulin receptor and cytokines such as TNFα and IL-6 can be considered as the molecular basis of organ crosstalk. This aspect will be considered in detail in Chapter 5. After the activation of the insulin receptor tyrosine kinase by autophosphorylation of its activation loop, the kinase phosphorylates tyrosine residues outside the kinase domain of the receptor, which creates binding signatures for signaling protein partners containing SH2 (src-homology 2) or PTB (phosphotyrosine-binding) domains. The insulin receptor does not bind signaling proteins directly but instead binds to a family of large docking proteins called IRS (insulin receptor substrate)1−6, as well as the adapter Shc (Src-homology 2 domain containing). This initial protein complex is of key importance for mediating the host of different downstream signaling pathways and final biological readouts addressed by insulin. For reviews on this topic, see White (2003) and Taniguchi et al. (2006). It is now widely accepted that IRS-1 and -2 are responsible for the vast majority of the effects of insulin, whereas other docking proteins such as CBL play a minor and much less investigated role. IRS proteins contain up to 20 potential Tyr phosphorylation sites that are able to bind to SH2 domains found in a variety of signaling proteins. This explains the complicated downstream diversification of the insulin signal. The two main pathways of insulin signaling originating from the IRS level are the phosphatidylinositol 3-kinase (PI3K, a lipid kinase)/AKT (also known as PKB or protein kinase B) pathway and the Raf/Ras/MEK MAPK (mitogen-activated protein kinase, also known as ERK or extracellular signal−regulated kinase) pathway. The PI3K pathway is to a large part mediating the metabolic effects of insulin and is connected exclusively through IRS. Remarkably, the MAPK pathway originates from both IRS and Shc and is involved in the regulation of gene expression and, in cooperation with the PI3K pathway, in the control of cell growth and differentiation (Taniguchi et al., 2006).

The prototypical metabolic effect of insulin is the stimulation of glucose transport in AT and skeletal and cardiac muscle. Given that skeletal muscle is one of the largest organs in the body, glucose disposal into muscle can be regarded as the major component of insulin action that prevents postprandial hyperglycemia. This effect is achieved through the translocation of the insulin-sensitive glucose transporter GLUT4 from intracellular vesicles to the plasma membrane. This process still remains only partly understood. It is also noteworthy that muscle contraction, and hence exercise, is able to induce GLUT4 translocation in a completely insulin-independent fashion (Huang and Czech, 2007). GLUT4 is one of 13 human glucose transporter isoforms (GLUTs) with 12 membrane-spanning domains and is expressed preferentially in AT and skeletal muscle. GLUTs mediate hexose transport across cell membranes through a so-called facilitative diffusion mechanism that is different from active, energy-consuming transport systems. In contrast to GLUT1,

GLUT4 has the unique characteristic of a mostly intracellular disposition in the unstimulated state. It is found in highly specific storage vesicles that are redistributed to the plasma membrane in response to insulin. The major insulin signaling pathway involved in GLUT4 translocation is the PI3K/PDK1/AKT2 pathway, resulting in phosphorylation of the Akt substrate AS160. AS160 is a GTPase-activating protein that when phosphorylated activates small G-proteins called RAB-GTPases that play multiple roles in membrane trafficking, by blocking the exchange of GTP for GDP. This tremendously detailed knowledge on the molecular steps of insulin action has paved the way to a much deeper understanding of the mechanisms underlying the crosstalk, specifically the negative crosstalk between AT and skeletal muscle.

In fact, increased AT mass and insulin resistance are closely related. In metabolic diseases, skeletal muscle insulin resistance results from an increased release of proinflammatory adipokines, cytokines, chemokines, and free fatty acids (FFAs) as a result of AT dysfunction. In a pioneer coculture model with human adipocytes and myocytes, our group demonstrated that adipocytes directly induced insulin resistance in human skeletal muscle cells (Dietze et al., 2002). The coculture reduced insulin-mediated Akt phosphorylation, GLUT4 translocation, and glucose uptake (Dietze et al., 2002) in a similar way to what was reported in skeletal muscle of diabetic patients (Krook et al., 2000). CM from adipocytes increased oxidative stress and ceramide content, reduced mitochondrial capacity, and potentiated the lipotoxic potential of palmitate in skeletal muscle cells (Sell et al., 2006a). These findings underpin the crucial role of adipokines in skeletal muscle dysfunction and in the pathogenesis of type 2 diabetes.

TNFα AND IL-6

TNFα is a 25 kDa proinflammatory protein highly expressed in monocytes and macrophages, playing a major role in inflammation and immunomodulatory processes. TNFα can be considered as a paradigm adipocytokine, setting the stage for a new understanding of the endocrine function of AT (Hotamisligil et al., 1993) originally obtained in rodent models. In humans, TNFα is not released from adipocytes; however, on development of obesity, macrophages infiltrate into AT representing a major source of TNFα released to the circulation. Early in vitro studies showed that TNFα induces insulin resistance in insulin-sensitive tissues (Cawthorn and Sethi, 2008), raising the hope that neutralization of TNFα might protect against obesity-induced insulin resistance. This was in fact demonstrated in a number of animal studies using either genetic ablation of the TNFα gene or treatment with TNFα antibodies (Uysal et al., 1997). Unfortunately, this concept could not be translated to the human setting, and despite chronic TNFα neutralization, no improvement of insulin sensitivity was found in insulin-resistant obese subjects (Fasshauer and Bluher, 2015). However, TNFα may exert important functions in the intraorgan crosstalk in human AT, an area that is still underexplored at the moment.

IL-6 is primarily known as a cytokine with a strong relation to inflammatory diseases and the development of insulin resistance in different target tissues of insulin (Catoire and Kersten, 2015). As described in the next chapter (Chapter 3), IL-6 gained considerable interest as the first and most studied myokine, most likely exerting beneficial effects. Given that IL-6 is also produced by AT, it can be considered as a paradigm of a so-called adipomyokine (separately addressed in Chapter 4). Chronically elevated IL-6 concentrations, as seen in obese subjects, may induce skeletal muscle insulin resistance, whereas acute elevation of this molecule can act as insulin-sensitizing agent (Weigert et al., 2005). A more detailed discussion of this rather interesting property will be presented in Chapter 4.

MCP-1 AND VASPIN

MCP-1 has originally been described as a secretory product of monocytes and endothelial cells with a role in atherosclerosis but was also allocated to the list of adipokines about 15 years ago with functional relevance for insulin resistance in adipocytes and myocytes (Sartipy and Loskutoff, 2003; Sell et al., 2006b). More recent studies have confirmed that adipocytes are an important source of MCP-1 and that it can induce AT inflammation even in the absence of macrophages (Sindhu et al., 2015). As a chemokine, MCP-1 is primarily involved in recruiting leukocytes into an area of inflammation, involving interaction of MCP-1 with its receptor CCR2. However, downstream signaling of the CCR2 receptor can also interfere with the insulin signaling cascade, as shown for human skeletal muscle cells (Sell et al., 2006b). In vivo, MCP-1 may act in a more indirect way by recruiting monocytes into different target tissues, which then release proinflammatory cytokines such as TNFα and IL-6, augmenting the negative crosstalk with the insulin signaling machinery.

When first described as a novel adipokine, the 47 kDa protein vaspin raised considerable interest, as it was found to be highly expressed in visceral AT and upregulated in obesity (Hida et al., 2005). In this early study, vaspin was described as a unique insulin-sensitizing adipokine. Subsequent studies demonstrated the expression of vaspin in additional tissues including skin, hypothalamus, and pancreatic islets. Importantly, application of recombinant vaspin to obese mice was found to improve insulin sensitivity and glucose tolerance (Youn et al., 2008). Recent data suggest that this effect of vaspin is due to the inhibition of kallikrein 7, a protease that has been found to be able to degrade insulin (Heiker et al., 2013). Thus, it is possible that vaspin augments the plasma insulin level leading to improved glucose metabolism.

ADIPONECTIN AND LEPTIN

The circulating levels of adiponectin are downregulated in obesity and are inversely correlated to insulin resistance. It is also well established that adiponectin improves whole body insulin sensitivity with a prominent effect on skeletal muscle (Turer and Scherer, 2012). Interestingly, this adipokine has also been reported as a myokine released from L6 muscle cells (Liu et al., 2009). This may have some important autocrine/paracrine function, but the vast majority of adiponectin is released from adipocytes. Adiponectin directly prevented insulin resistance in myocytes cocultured with adipocytes, but this effect was shown to be due to the control of the adipocyte secretome by adiponectin (Dietze-Schroeder et al., 2005). Another proposed mechanism is that the globular C-terminal fragment of adiponectin can reduce glucose levels by increasing fatty acid oxidation in myocytes (as reviewed in Turer and Scherer, 2012). It was demonstrated that decreased levels of adiponectin and the adiponectin receptor AdipoR1 in obesity may have causal roles in mitochondrial dysfunction and insulin resistance observed in diabetes. Adiponectin increased PGC-1α expression and mitochondrial content in myocytes via the AdipoR1 receptor. Analogously, a transgenic mouse with skeletal muscle—specific disruption of AdipoR1 displayed a decreased expression of PGC-1α, mitochondrial content, and detoxifying enzymes in skeletal muscle, which were associated with insulin resistance and decreased exercise endurance in these mice (Iwabu et al., 2010). Most of the reported effects of adiponectin in skeletal muscle were obtained using globular adiponectin, which may have a reduced contribution to overall circulating adiponectin. Therefore, the real physiological impact of adiponectin on skeletal muscle needs to be further explored.

Leptin circulating levels positively correlate with fat mass. Leptin has been proposed to be an adipomyokine, as it is secreted from both adipocytes and myocytes. However, data from our group showed that leptin is rather a true adipokine (Raschke and Eckel, 2013). Leptin receptors are abundant in human skeletal muscle, and their expression is upregulated by both exercise and atrophy (Chen et al., 2007). Leptin promotes insulin sensitivity by enhancing fatty acid oxidation and decreasing triglyceride storage in muscle (Ahima and Flier, 2000). Several studies with murine models support an anabolic role for leptin in the regulation of muscle mass. A novel mouse model lacking all functional leptin receptor isoforms, the POUND mouse (Lepr[db/lb]), showed increased body weight and decreased muscle mass. In POUND mice, myogenic differentiation was probably impaired owing to an increase in muscle-wasting myostatin and decreased IGF-1 and Akt expression in skeletal muscle (Arounleut et al., 2013). It was also shown that leptin treatment increases skeletal muscle mass in aged mice. Thus, future research has to determine whether leptin could be a therapeutic target to improve declined skeletal muscle function in degenerative diseases and aging.

DPP4 AND VISFATIN IMPACT ON SKELETAL MUSCLE

Our group has characterized DPP4 as an adipokine (Lamers et al., 2011) potentially linking obesity to the metabolic syndrome. In mice lacking DPP4, both insulin secretion and glucose tolerance are improved (Marguet et al., 2000). In a recent study, DPP4 inhibition with sitagliptin was shown to upregulate GLUT4 expression in skeletal muscle of spontaneous hypertensive rats (Giannocco et al., 2013). Nevertheless, the impact of adipose-derived DPP4 on skeletal muscle remains fully unknown.

Visfatin was originally described as an adipokine with insulin-mimetic properties, although this later affirmation was retracted (Fukuhara et al., 2007). Visfatin is identical to the cytokine pre-B cell colony-enhancing factor and the enzyme nicotinamide phosphoribosyltransferase (Nampt), catalyzing a rate-limiting step in the synthesis of nicotinamide adenine dinucleotide (NAD). Little is known on the potential impact of this adipokine on skeletal muscle. Visfatin may have a deleterious effect on skeletal muscle because it induced oxidative stress in C2C12 myotubes through nuclear factor (NF)-κB activation independently of MAPK and Akt phosphorylation (Oita et al., 2010). Costford et al. have demonstrated that exercise increases intracellular Nampt (iNampt) expression in human skeletal muscle and that iNampt induction correlates with an enhancement of mitochondrial content and PGC-1α expression (Costford et al., 2010). Future work will be required to better define the role and the specific impact of visfatin for the fat—muscle crosstalk axis.

THE FAT—LIVER CROSSTALK

In addition to skeletal muscle, the liver plays a key role in systemic glucose regulation and is essentially involved in lipid homeostasis. In the fasting state, more than 90% of endogenous glucose production arises from the liver, and increased fasting hyperglycemia in type 2 diabetes is mostly related to increased hepatic gluconeogenesis (Cusi, 2010). It is generally accepted that an increased level of circulating FFAs leads to hepatic insulin resistance, impaired insulin signaling, and increased hepatic glucose production (Roden et al., 2000). Importantly, lipid accumulation by the liver leads to nonalcoholic fatty liver disease (NAFLD), a very common disorder in patients with type 2 diabetes being frequently associated with hepatic insulin resistance (Cusi, 2009). In obese subjects, the prevalence of NAFLD can be as high as 90%. Although lipid flux to the liver is a major cause of NAFLD, adipokines are importantly linked to this disorder, potentially representing new targets for the treatment of NAFLD (Polyzos et al., 2016). A short overview of the most recent findings regarding this scenario is presented below. A graphical overview of the most prominent hepatic effects of key adipokines is presented in Fig. 2.5.

Adiponectin: As described before, adiponectin is one of the most abundant proteins produced by mature adipocytes with prominent effects on the liver including an improvement of insulin resistance, hepatic steatosis, inflammation, and fibrosis. The biological effects of adiponectin are exerted by two transmembrane receptors, AdipoR1 and AdipoR2, (Yamauchi et al., 2003) that are both expressed in the liver. Adiponectin decreases gluconeogenesis in the liver, most likely by reducing the expression of several gluconeogenic enzymes such as PEPCK and glucose-6-phosphatase (Yamauchi et al., 2002). Earlier studies suggested that this could be due to activation of the AMPK pathway by adiponectin. However, more recent data show that the reduction of gluconeogenic gene expression is independent of AMPK (Miller et al., 2011), and additional pathways potentially involving both adiponectin receptors appear to be involved. Adiponectin is also able to prevent hepatocytes from apoptosis, a process with deep impact for NAFLD. Adiponectin also acts on nonhepatocytes such as Kupffer cells and exerts a prominent

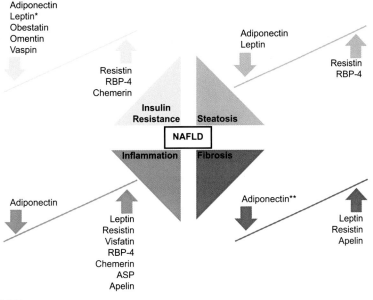

FIGURE 2.5

Hepatic effects of adipokines. *NAFLD*, nonalcoholic fatty liver disease; *RBP4*, retinol-binding protein 4.

*Leptin improves insulin resistance under normal conditions.

**Adiponectin decreases hepatic fibrosis.

Reproduced with permission from Polyzos, S.A., Kountouras, J., Mantzoros, C.S., 2016. Adipokines in nonalcoholic fatty liver disease, Metabolism 65, 1062–1079.

antiinflammatory action by reducing the production of cytokines such as IL-6 and TNFα from these cells. Overall, adiponectin is a paradigm molecule involved in a protective crosstalk from fat to the liver.

Leptin: Similar to adiponectin, leptin is known to improve hepatic insulin resistance and steatosis, most likely owing to the prevention of lipogenesis and the activation of fatty acid oxidation involving phosphorylation of ACC-1 (Huang et al., 2006). Undoubtedly, the liver must be considered as a major target of leptin, as evidenced by the abundant expression of leptin receptors in this tissue (Cohen et al., 2005). Downstream signaling of leptin in the liver involves the JAK/STAT3 pathway followed by activation of MAPK/ERK, STAT5, and transcriptional target networks (Robertson et al., 2008). However, leptin signaling is rather complex and additional pathways such as PI3K/Akt, AMPK, and FoxO1 may also be involved. Interestingly, there are several reports on a potential interplay between leptin and insulin signaling in the liver, but the precise mechanisms and the potential therapeutic implications of this signaling crosstalk require further investigations (Polyzos et al., 2016). Recently, a role for leptin in the incorporation of triglycerides into VLDL was demonstrated by generating a liver-specific knockout of leptin receptors (Huynh et al., 2013). Thus, leptin may be of importance for controlling hepatic lipid export to other target tissues. In contrast to these beneficial effects on the liver, at least in animal models, leptin was found to promote hepatic inflammation and fibrosis. This involves a variety of pathways including upregulation of collagen type 1 and the repression of MMP-1, a key factor in tissue remodeling. Recent data show that leptin is able to upregulate miR21 and fibrogenesis involving activation of TGF-β signaling (Dattaroy et al., 2015). It can be concluded that leptin deficiency is implicated in hepatic steatosis but not inflammation, whereas high leptin levels may lead to inflammation and fibrosis. It is, however, important to note that this potentially dual action of leptin has not been validated in humans and that the clinical application of leptin and leptin analogs is still of great interest, specifically for treatment of lipodystrophy and congenital leptin deficiency. As pointed out by Mantzoros and coworkers (Polyzos et al., 2016), this will require strictly controlled studies in well-defined cohorts in the future. *Other adipokines:* As shown in Fig. 2.5, a number of additional adipokines such as resistin, chemerin, visfatin, apelin, and many others may play a role in the crosstalk from AT to the liver. Compared with leptin and adiponectin, this scenario has been much less explored, and the majority of these molecules do play a role under pathophysiological conditions. Thus, with ongoing AT inflammation and macrophage infiltration, these molecules are hypersecreted and able to negatively affect liver function. A detailed description of this topic will be presented in Chapter 5.

THE FAT–PANCREAS CROSSTALK

Both pancreatic α- and β-cells are key regulators of glucose homeostasis by secreting glucagon and insulin. This process is directly controlled by glucose involving an increase in cellular ATP, changes in membrane polarity leading to differential regulation of Ca^{++} fluxes, with subsequent inhibition of glucagon release in α-cells, and stimulation of insulin secretion from β-cells. Although glucose can be considered as the fundamental regulator of the secretory function of these pancreatic cells, recent data have shown an important impact of various adipokines on this process, and the fat–pancreas axis has gained considerable interest, specifically in obesity and type 2 diabetes.

Impact on α-Cells

It is known for longtime that leptin suppresses glucagon secretion, as seen in mouse models, with a defect in leptin signaling that develop hyperglucagonemia (Dunbar and Walsh, 1980). Recent studies have shown that pancreatic α-cells express leptin receptors, and leptin was found to decrease glucagon secretion, at least under in vitro conditions. However, this effect is much less prominent under in vivo conditions and may involve (Soedling et al., 2015) central actions of leptin. Additional future work needs to address this important issue. Similarly, the role of adiponectin for the secretory function of pancreatic α-cells remains unexplored, although this may represent an important target of adiponectin and may be of relevance for glucose homeostasis, under both normal and pathophysiological conditions.

Impact on β-Cells

Type 2 diabetes results from pancreatic β-cell failure to secrete enough insulin to compensate for increasing insulin resistance. In early stages of the disease, insulin resistance is compensated by increasing β-cell function and mass. Over disease progression, there is an increased deterioration of both β-cell function and mass, mainly due to accelerated apoptosis. The main factors contributing to this loss of β-cell function and mass are glucotoxicity, lipotoxicity, and islet cell amyloid (Wajchenberg, 2007). Thus, the maintenance of β-cell function and mass are crucial targets to prevent the onset of type 2 diabetes.

It is now generally accepted that the proinflammatory adipokine profile in obese AT contributes to pancreatic β-cell injury. While the induction of β-cell apoptosis by IL-1, TNFα, and IFN-γ (interferon γ) is well established, other rather adipose-specific adipokines such as leptin, resistin, and adiponectin have been proposed as important players in the development of pancreatic β-cell dysfunction and type 2 diabetes. In the following part, the main direct effects of the classical adipokines, adiponectin and leptin, and the novel adipokines, visfatin and DPP4, will be discussed.

ADIPONECTIN AND LEPTIN: ANTAGONIST EFFECTS ON PANCREATIC β-CELLS

The adiponectin receptors AdipoR1 and AdipoR2 are expressed in primary and clonal β-cells (Wijesekara et al., 2010). The C-terminal globular domain of adiponectin inhibited the apoptotic effect of cytokines and palmitate and prevented the impairment of insulin secretion induced by FFA and cytokines in INS-1 rat clonal cell line (Rakatzi et al., 2004). Similarly, adiponectin reversed apoptosis triggered by both intermittent and sustained high glucose in INS-1 cells. However, adiponectin did not inhibit FFA-induced apoptosis in isolated human islets (Staiger et al., 2005). In isolated islets from mice under high-fat diet, adiponectin stimulates glucose-stimulated insulin secretion (GSIS). Furthermore, several studies have provided evidence that adiponectin stimulates insulin secretion through direct exocytosis of insulin granules and upregulation of insulin gene expression. It has recently been proposed that adiponectin's beneficial effects on β-cell survival rely on ceramidase activity leading to formation of the antiapoptotic metabolite sphingosine-1-phosphate (S1P) under both in vitro and in vivo conditions (Holland et al., 2011). All this bulk of evidence supports the notion that adiponectin positively affects β-cell function and growth.

Leptin was the first adipokine suggested to directly exert pancreatic effects. In addition to its central action on food intake and energy expenditure, leptin has been proposed to regulate glucose homeostasis by modulating the synthesis and release of insulin and glucagon through direct actions on β- and α-cells (see above) of pancreatic islets, respectively (Dunmore and Brown, 2013). The leptin receptor ObR is highly expressed in the endocrine pancreas, and clinical data suggested a negative correlation between leptin circulating concentrations and β-cell function (Ahren and Larsson, 1997). In most studies with perfused rat pancreas and rodent islets, physiological concentrations of leptin-induced GSIS from pancreatic β-cells (as reviewed in Seufert, 2004). Furthermore, leptin suppressed insulin gene expression and GSIS. On the other hand, leptin has been shown to inhibit ectopic fat accumulation, to prevent β-cell dysfunction, and to protect the β-cell from cytokine- and fatty acid−induced apoptosis (Brown and Dunmore, 2007). The impaired leptin signaling in the leptin-deficient mouse (*ob/ob*) and the leptin receptor−deficient mouse (*db/db*) leads to hyperinsulinemia even before the development of the obese and diabetic phenotype (Coleman, 1982). Although initial studies provided conflicting results on the effects of leptin on insulin secretion, probably owing to the U-shaped effect depending on the leptin concentrations, it is currently accepted that leptin inhibits insulin secretion in β-cells both in vitro and in vivo (as reviewed in Dunmore and Brown, 2013).

DPP4 AND VISFATIN: POTENTIAL MEDIATORS OF β-CELL DYSFUNCTION?

The direct effects of DPP4 on β-cell function remain unclear, but owing to its enzymatic cleavage of GLP1, this novel adipokine may have a crucial effect on β-cells. GLP1 has a prominent role in β-cell function, as it is responsible for approximately 70% of postprandial insulin secretion. GLP-1 also contributes to glucose homeostasis through its effects on insulin biosynthesis and its inhibition of glucagon release. Furthermore, GLP1 promotes pancreatic β-cell proliferation. The beneficial effects of GLP-1 on pancreatic β-cells have also been reported in studies with diabetic animals (as reviewed in Garber, 2011). Similarly, DPP4 pharmacological inhibition has been related to improved β-cell survival and neogenesis in streptozotocin-treated diabetic rats. The DPP4 inhibitors sitagliptin, saxagliptin, and vildagliptin improved β-cell function as proinsulin-to-insulin ratios decreased and HOMA-B increased in type 2 diabetic patients (as reviewed in Cernea and Raz, 2011). In the light of these evidences, this novel adipokine may have negative effects on the pancreatic β-cell counteracting the positive effects of GLP1 on pancreatic β-cell mass and glucose homeostasis.

Visfatin has intrinsic enzymatic activity as Nampt (Revollo et al., 2007). Visfatin can directly increase insulin secretion in β- and βTC6 cells (Revollo et al., 2007). Visfatin effects on β-cells may be mediated by its intrinsic enzymatic activity, as visfatin effects on insulin secretion can be blocked by FK866; the enzymatic inhibitor of Nampt and the enzymatic product of visfatin, NMN, have been proposed to regulate insulin transcription through PDX1 expression. In MIN6 cells, visfatin stimulated β-cell proliferation and inhibited palmitate-induced β-cell apoptosis through ERK1/2 and PI3K/Akt activation (Cheng et al., 2011). In contrast, Nampt and NMN did not have an effect on β-cell survival but rather both potentiated GSIS in human islets in a recent study (Spinnler et al., 2013). Nampt heterozygous female mice (Nampt+/−) showed impaired glucose tolerance owing to reduced GSIS (Revollo et al., 2007). NMN administration reverted impairment of β-cell function and prevented the inflammation-induced islet dysfunction in mice under a fructose-rich diet (Caton et al., 2011). These potential protective effects are in conflict with one clinical study that correlated visfatin circulating levels with progressive β-cell deterioration (Lopez-Bermejo et al., 2006). It has been speculated that visfatin's beneficial effects on β cell function occur at lower physiological levels, whereas higher concentrations within the pathophysiological range may have harmful effects. All these studies underpin the relevant role of visfatin through its Nampt enzymatic activity regulating β-cell function.

The Adipovascular Axis

Vascular complications are the main cause for mortality in obese and type 2 diabetic patients. Indeed, obesity is an independent risk factor for cardiovascular (CV) atherosclerosis (Williams et al., 2002). Endothelial dysfunction is a key initial step in the development of atherosclerosis, which is characterized by reduced bioavailability of the antiatherogenic molecule, nitric oxide (NO), and impaired vascular homeostasis leading to increased vasoconstriction, leukocyte adherence, platelet activation, smooth muscle proliferation, permeability, prooxidation, coagulation, and inflammation (Verma et al., 2003). Atherogenesis is a complex process beginning with endothelial activation. The increased expression of adhesion molecules promotes leukocyte adhesion and transmigration. After lipid accumulation, activated monocytes become foam cells. Later they are joined by smooth muscle cells that migrate and proliferate toward the intima layer and increase extracellular matrix secretion and the release of proinflammatory cytokines leading to fibrous cap formation. When the fibrous cap weakens by the action of MMPs, the plaque becomes unstable and breaks leading to thrombus formation and ischemic accident. NF-κB plays a pivotal role in atherothrombotic diseases, as it regulates the expression of other key molecules in this process such as adhesion molecules, proinflammatory cytokines, and the inducible form of NO synthase (iNOS), a key enzyme related to impaired vascular reactivity in diabetes (Gunnett et al., 2003). A proinflammatory secretion profile of AT has been associated with chronic systemic inflammation and endothelial dysfunction. Adipocytokines such as TNFα, several interleukins, and many other adipokines lead to the activation of proinflammatory signaling pathways in vascular cells. AT can additionally affect vascular homeostasis through the release of classical vasoactive factors such as NO or angiotensin II, agents controlling fibrinolysis such as PAI-1 (plasminogen activator inhibitor-1), and vasoactive adipokines (Guzik et al., 2006). Several adipokines are now considered as markers or predictors of CV disease (Taube et al., 2012), but more importantly, adipokines are starting to be acknowledged as active promoters of atherothrombotic diseases. The main direct CV effects of adiponectin, leptin, DPP4, and visfatin will be addressed here.

ADIPONECTIN AND LEPTIN: TWO OPPOSITE PLAYERS IN THE VASCULAR WALL?

Considering the cardioprotective adipokine *par excellence*, adiponectin circulating levels are downregulated in patients suffering from cardiometabolic diseases

(Kumada et al., 2003). This adipokine inhibits the initial steps of the atherogenic process by downregulating TNFα induction of adhesion molecules, preventing macrophage-to-foam cell transformation and smooth muscle cell proliferation (Arita et al., 2002). This adipokine prevents endothelial activation by inhibiting NF-κB signaling and directly ameliorating endothelial dysfunction by increasing nitric oxide (NO) production (as reviewed in Taube et al., 2012). The antiatherogenic effect of adiponectin has been further demonstrated in adiponectin knockout models (Ouedraogo et al., 2007).

Leptin receptors are expressed in both endothelial cells and smooth muscle cells and thus can exert direct actions on the vascular wall. The correlation between leptin levels and CV diseases remains controversial; therefore, it has been proposed that the positive or negative correlations reported may depend on the absence or presence of the specific CV pathophysiology (Sweeney, 2010). Leptin also increases the expression of PAI-1, CRP, and caveolin-1 in human coronary artery endothelial cells (Singh et al. 2007, 2010, 2011). Within physiological concentrations, leptin promotes the release of vasodilator factors such as NO and endothelium-derived hyperpolarizing factor (Beltowski, 2012), whereas at concentrations in the pathophysiological range (hyperleptinemia), leptin impairs NO-mediated vasodilation by acetylcholine both in vitro and in vivo (Knudson et al., 2005). However, the in vivo effects of leptin on atherogenesis remain elusive. Leptin-deficient hyperlipidemic mice develop less atherosclerosis than the proatherogenic apoE$^{-/-}$ mice on an atherogenic diet. On the contrary, low-density lipoprotein−receptor knockout mice (LDLR$^{-/-}$) lacking leptin develop more atherosclerotic lesions than LDLR$^{-/-}$ control mice. The conflicting results from in vivo clinical and animal studies demonstrate that the contribution of leptin to obesity-related CV complications is complex and may depend on the pathological scenario, the degree of hyperleptinemia, and the progression of the CV disease.

DPP4 AND VISFATIN: PROMISING THERAPEUTIC TARGETS IN CARDIOVASCULAR DISEASES

DPP4 is an exoprotease that cleaves and inactivates several substrates such as the members of the incretin family, glucagon-like peptide-1 (GLP-1) and gastric inhibitory polypeptide (Drucker and Nauck, 2006). The soluble form of DPP4 displays enzymatic activity and is found in plasma (Iwaki-Egawa et al., 1998), but its direct effects on the vascular wall remain unknown. Our group demonstrated that DPP4 directly induces proliferation of human vascular smooth muscle cells (Lamers et al., 2011). DPP4 increases superoxide generation and the receptor of advanced glycation end products (AGEs) gene expression in HUVECs in a concentration-dependent manner (Ishibashi et al., 2013). AGEs are products of nonenzymatic glycation and oxidation of proteins and lipids that accumulate in hyperglycemia. Thus, both AGEs and its receptor RAGE play a key role in diabetic-related vascular complications. Both DPP4 pharmacological inhibition and genetic deletion have been proposed to upregulate cell survival pathways in cardiomyocytes and endothelial cells through a GLP-1—dependent mechanism (Bose et al., 2005). However, increasing evidence suggest that DPP4 inhibitors exert vasoprotective effects independent of GLP-1 through endothelial repair, antiinflammatory effects, and inhibition of ischemic injury (Fadini and Avogaro, 2013). Thus, DPP4 inhibition delays the onset of atherosclerosis in a murine model of atherosclerosis and insulin resistance (Shah et al., 2011). On the contrary, it has been described that loss of DPP4 enzymatic activity in HUVEC can induce a prothrombotic status (Krijnen et al., 2012). Although DPP4 inhibition represents a promising therapeutic approach, further research on the direct action of DPP4 on the vascular wall is required.

Visfatin circulating levels are increased in obesity, type 2 diabetes mellitus, and atherothrombotic diseases. This adipokine promotes angiogenesis and smooth muscle cell proliferation and triggers endothelial cell proliferation and migration. Furthermore, in the latter cell type, visfatin upregulates adhesion molecules, cytokine secretion, and NADPH oxidase activation via NF-κB, whereas in vascular smooth muscle cells, it activates the ERK1/2-NF-κB—iNOS axis. Visfatin promotes atheroma plaque destabilization through MMP-2 and 9 in both endothelial cells and monocytes. Regarding effects on the vascular tone, visfatin induces relaxation in rat aortic rings, whereas in bovine coronary arteries and human and murine microvessels, visfatin impairs endothelium-dependent relaxation to ACh (as reviewed in Romacho et al., 2013). On the contrary, visfatin has an antiapoptotic effect on cardiomyocytes, and its proangiogenic properties are beneficial in peripheral limb ischemia. Both clinical and basic research evidence supports a role for visfatin as a promising therapeutic target in metabolic-related CV diseases.

Additional Fat Depots With a Specific Role in Organ Crosstalk

In addition to the subcutaneous and visceral fat depot, a number of smaller locations of AT exist with highly specific functions. (For recent reviews on this topic, please see Carobbio et al., 2017; Contreras et al., 2017b; Crewe et al., 2017; Kiefer, 2017; Lynes and Tseng, 2017; Murawska-Cialowicz, 2017; Schafer et al., 2017.) Some of these depots such as perivascular and epicardial fat are in direct touch with other tissues and exert more local functions such as substrate supply, mechanical protection, and others. In contrast, BAT, although present in rather small quantities in humans, is able to substantially affect metabolic control of the whole body, and this issue has raised considerable interest. The following sections focus on the secretome of these adipose depots and the differentiation compared with that of the classical WAT. At present, our knowledge on this topic is rather limited and future studies are needed to fully delineate the crosstalk scenario involving these specific fat depots.

PERIVASCULAR ADIPOSE TISSUE

Obesity is a key driver of the metabolic syndrome, a complex entity of several risk factors being tightly associated with CV disease and other vascular abnormalities. This has raised considerable interest in the biology of perivascular AT as a unique fat compartment and its impact for vascular dysfunction beyond the effects of adipocytokines released from visceral and/or subcutaneous fat described earlier. A number of excellent reviews on this topic are recommended for further reading (Ayala-Lopez and Watts, 2017; Nosalski and Guzik, 2017; Ramirez et al., 2017; Schafer et al., 2017; Xia and Li, 2017; Zaborska et al., 2017).

Perivascular AT is a paradigm of a locally acting fat depot, in close contact (without any barrier) with the adventitia of large, medium, and small diameter arteries. Perivascular adipocytes are thought to signal locally in a para- and vaso-crine fashion. Classically, perivascular fat was thought to have only mechanical properties, but it turned out that this fat depot is metabolically very active, and in addition to the local effects, perivascular fat also exerts endocrine functions and communicates with other organs, such as the liver (Siegel-Axel and Haring, 2016). When compared with the total body fat, perivascular AT turns out to be a very small depot amounting to only 3% of total AT. However, as the vast majority of systemic arteries and small vessels are surrounded by perivascular fat, this depot merits specific attention. With respect to different vascular beds, different functional properties of perivascular fat were reported (Gil-Ortega et al., 2015).

A prominent physiological function of perivascular fat relates to the anticontractile or vasodilative effects exerted by this tissue (Gollasch, 2012). Although postulated more than 30 years ago, only recently several studies addressed the identification of a so-called adipocyte-derived relaxing factor (Dubrovska et al., 2004; Fesus et al., 2007; Gollasch, 2012; Gollasch and Dubrovska, 2004; Nava and Llorens, 2016). At present, several candidates such as adiponectin, NO, prostacyclin, and angiotensin are considered as mediators of vascular relaxation, but additional adipokines could be involved and remain to be identified.

Perivascular fat plays an important role for the CV system, and recent reports have highlighted that human coronary artery perivascular adipocytes are responsible for regulating vascular morphology, inflammation, and homeostasis (Chatterjee et al., 2013). Inflammation induced by factors released from enlarged perivascular fat around coronary arteries induces progression of atherosclerosis. Specifically, IL-8, IL-1, and MCP-1 were found to be released by perivascular AT (Verhagen and Visseren, 2011). These cytokines are key mediators of inflammation and invasion of immune cells into the vessel wall, contributing to the progression of atherosclerosis and plaque vulnerability. An additional player involved in cardiac and CV dysfunction is the so-called epicardial fat (see below). However, epicardial fat does not directly surround coronary arteries; instead, it is in direct contact with cardiomyocytes. Furthermore, some data indicate that coronary perivascular fat is metabolically more active than epicardial AT (Spiroglou et al., 2010). Considering this, an important question relates to the potential differences between subcutaneous, visceral, and perivascular fat in terms of the expression and secretory signature. In an elegant study, Rittig et al. (2012) analyzed the mRNA expression and protein secretion of perivascular fat cells, which were obtained from fat depots located around human arm arteries. In this study, they could show that perivascular (pre-) adipocytes differ substantially from subcutaneous and visceral (pre-) adipocytes in terms of mRNA and protein expression of inflammatory, metabolic, and angiogenic factors. Compared with subcutaneous or visceral adipocytes, perivascular fat cells secrete a number of angiogenic proteins such as HGF-1, VEGF, TSP-1, and IGFBP-3 at a much higher rate. This supports the notion that perivascular AT plays an important role for arterial vessel wall physiology and that secretory malfunction (see Chapter 5) of this fat depot is critically involved in different aspects of atherosclerosis such as plaque growth and differentiation.

Owing to its specific anatomic location, perivascular AT primarily exerts a paracrine exchange of secretory products between vascular cells including adventitial fibroblasts, medial smooth muscle cells, and intimal endothelial cells (Lee et al., 2009). However, owing to infiltration of perivascular fat cells into the adventitia a close contact to vasa vasorum may exist. This leads to diffusion of secretory products to the blood stream, and in that way, perivascular AT may also act as an endocrine organ. The precise contribution to an endocrine crosstalk is presently unclear, mostly owing to the very limited number of studies on primary human perivascular fat cells. Importantly, it was shown that perivascular fat cells both in vitro and in vivo have a much higher capacity to secrete angiogenic factors, specifically HGF. In a clinical study, HGF levels were measured and the tissue mass of different fat compartments was monitored by MRI. The data confirmed enhanced secretion of HGF in tight correlation with an increased perivascular fat mass (Rittig et al., 2012), providing strong evidence for an endocrine function of perivascular fat.

EPICARDIAL ADIPOSE TISSUE

As pointed out before, nearly all arteries and the heart are surrounded by fat depots. In the case of the human heart, two anatomically distinct fat depots cover approximately 80% of the heart, but the terminology used to describe these depots has been confusing. Here we use the term epicardial adipose tissue (EAT) when referring to the AT located between the myocardium and the visceral pericardium. The location of epicardial fat is illustrated in Fig. 2.6. The secretory function of epicardial fat has gained considerable interest, and a number of studies including analysis of human epicardial fat have been conducted in recent years. Excellent reviews on this topic have been published and are recommended for further details on this developing topic (Chistiakov et al., 2017; Iacobellis, 2016; Matloch et al., 2016; Patel et al., 2017; Rietdorf and MacQueen, 2017; Salazar et al., 2016; Wang et al., 2016; Wu et al., 2017).

Epicardial fat cells have the same embryologic origin as mesenteric and omental fat cells and are derived from the splanchnopleuric mesoderm associated with the gut. EAT constitutes approximately 20% of the total ventricular weight in the human heart and is commonly found at both ventricles in the atrioventricular and interventricular grooves extending to the apex and along the coronary arteries. EAT and the underlying myocardium share the same coronary blood supply, and no fascialike structure separates the two tissues. Based on these characteristics, EAT is considered to represent the true visceral fat depot of the heart. Compared with other fat depots, the number of adipocytes per gram of EAT is higher, and their size is smaller (Barber et al., 2000). Furthermore, differences in protein content and fatty acid

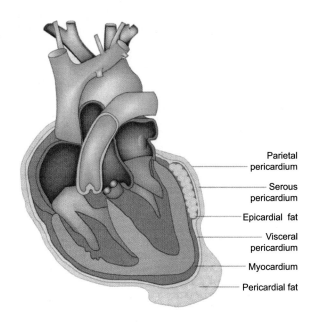

FIGURE 2.6

Location of epicardial fat.

Reproduced with permission from Iacobellis, G., 2016 Nat rev Endocrinol 11: 363–71

composition have been described for epicardial fat as compared with other fat depots in animal models. Importantly, the relative expression levels of adipokines in EAT differ from that of other fat depots in guinea pigs and humans (Fain et al., 2010; Swifka et al., 2008).

Studies toward the metabolic properties of EAT are limited. Mitochondrial copy numbers did not differ among various fat depots studied in guinea pigs, suggesting that the oxidative capacity of EAT is similar to that of other depots. EAT from guinea pigs also showed enhanced rates of basal fatty acid uptake, lipogenesis, and lipolysis when compared with the other fat depots studied. These findings were substantiated in human EAT by higher mRNA expression levels of the lipoprotein lipase and hormone-sensitive lipase as compared with subcutaneous AT (Fain et al., 2010).

PHYSIOLOGICAL FUNCTION OF EPICARDIAL ADIPOSE TISSUE

Physicians in the 18th and 19th century attributed cases of sudden death in corpulent subjects to the fatty heart, a condition where the heart is completely embedded in AT. These early observations are suggestive of an involvement of expanded EAT in the pathogenesis of CV disease. Under normal conditions, several functions have been proposed for EAT. Fatty acids represent the major energy source for the healthy heart to maintain contractile function (Lopaschuk, 2002). Therefore, the high rates of lipolysis observed in guinea pigs suggest that EAT could serve as local energy source for the heart. Based on the high rates of fatty acid uptake and synthesis by EAT observed in guinea pigs, it has been proposed that EAT might serve a protective role against elevated levels for FFAs in the coronary circulation (Marchington et al., 1989). In humans, mRNA levels of UCP1 and different transcriptional regulators are higher in EAT than in other (thoracic) fat depots (Sacks et al., 2009). This suggests that EAT could act like brown fat to protect the myocardium and coronary vessels against hypothermia. Finally, adipokines secreted from EAT, such as adiponectin, adrenomedullin, and omentin, may have protective effects on the myocardium and vasculature by regulating energy substrate and Ca^{2+}-metabolism (Greulich et al., 2013). It should be noted that experimental evidence supporting these functions is still limited. Most likely, this is due to the minute amounts of EAT present in small laboratory rodents such as mice and rats compared with those in larger mammals and humans.

CROSSTALK BETWEEN EPICARDIAL ADIPOSE TISSUE AND THE MYOCARDIUM

Adipokine expression in EAT has been profiled using microarray analysis and real-time PCR in humans and antibody arrays in guinea pigs (Greulich et al. 2011, 2012; Dutour et al., 2010). All studies show changes in adipokine expression levels among the various fat depots examined. Only a few studies have determined adipokine secretion from EAT. EAT explants were found to secrete IL1β, IL6, IL6 sR, MCP1, and TNFα, and specifically in explants from type 2 diabetic patients, an increased release of activin A, angiopoietin-2, and CD14 is observed (Greulich et al., 2012). Because EAT shares the coronary blood supply with the myocardium and no structures resembling a fascia separate the fat tissue and the myocardial layer, an active crosstalk between the fat cells and the cardiomyocytes may take place. However, the effects of adipokines are complex, and in addition to a dose-dependent setting, a certain combination of adipokines may induce the end-point effect. Thus, some adipokines such as omentin-1 and adiponectin are cardioprotective, some are deleterious such as resistin and activin A (Venteclef et al., 2015), and some even exhibit opposite effects depending on the concentration, such as leptin and apelin. An excellent and comprehensive overview of CV effects of adipocytokines secreted by EAT has been published in a recent review (Rietdorf and MacQueen, 2017). Greulich et al. (2012) were able to show that secretory products from EAT when applied to cardiomyocytes are able to induce a reduction in sarcomere shortening, cytosolic Ca^{++} fluxes, and a compromised insulin signaling. Activin A was identified as a major driver of these effects. These data support the notion that crosstalk from epicardial AT to cardiomyocytes is of key importance for cardiac function and performance.

BROWN ADIPOSE TISSUE

AT is traditionally classified as WAT and BAT. These different tissues have different morphology, biochemical features, and functions. WAT is the more abundant AT, representing at least 10% of the body weight of normal, healthy adult humans. As described earlier, it is primarily located in the subcutaneous and visceral depots and is able to store energy as triglycerides and to modulate energy homeostasis through its own secretory activity. Smaller depots with unique functions have been addressed earlier in this chapter. The white adipose cells are spherical with a variable size, which depends on the size of the lipid droplet (unilocular lipid droplet). In white adipocytes, the mitochondria, which are few in number and low in activity, appear thin and elongated and have few cristae. Conversely, BAT stores lipids in multilocular lipid droplets and is highly vascularized and innerved, in particular, by the sympathetic nervous system (SNS). In addition, brown fat cells themselves have a very unique cellular and molecular composition. The thermogenic property of brown fat cells is carried out by the very dense, large, spherical, and packed-with-laminar-cristae mitochondria, a typical feature of these cells. Indeed, brown fat cells possess the highest levels of mitochondria in mammalian organisms. Nevertheless, macroscopically it is difficult to establish a clear histological separation of both tissues. Many experimental evidences have shown the presence of both white and brown adipocytes in the same depot. Cinti and coworkers proposed the transdifferentiation theory to explain why white and brown adipocytes are present within the same depot (Cinti, 2009). According to this theory, particular physiologic conditions, such as cold exposure, are able to induce the differentiation of white adipocytes into brown cells. Furthermore, brown cells can be transformed in white adipocytes when the energy balance is positive and the adipose organ requires increased storage capacity (Cinti, 2009). Reports from different independent groups (Granneman et al., 2005) confirmed the transdifferentiation theory by analyzing the number of white and brown adipocytes and apoptosis markers. In two different strains of mice, they detected a reduction in the number of white cells and an increase in brown cell population, whereas no changes were observed in the total number of fat cells and apoptosis markers were absent in white cells. Nevertheless, subcutaneous and omental WAT could have distinct transdifferentiation abilities. In murine models, the overexpression of PRDM16 induced the formation of brownlike adipocytes only in subcutaneous WAT as compared with omental (Ohno et al., 2012). This scenario has opened the concept of "browning" of AT that seems to be dependent from the recruitment of brown progenitors and the transdifferentiation of white adipocytes in brown. Activation of brown fat leads to increased energy expenditure, reduced obesity, and lower plasma glucose and lipid levels and exerts a prominent beneficial effect on metabolic homeostasis. This property and the possibility of browning of white fat have raised a huge interest in this specific fat depot and its potential therapeutic value. A number of excellent reviews were published on this topic in recent years, and the interested reader will find detailed descriptions of the biology and function of brite adipose tissue and BAT in these papers: Aldiss et al. (2017); Betz and Enerback (2017); Buscemi et al. (2017); Chen et al. (2017a); Chu and Gawronska-Kozak (2017); Chu and Tao (2017); Cinti (2017); Contreras et al. (2017a); El Hadi et al. (2017); Fernandez-Quintela et al. (2017) and Flouris et al. (2017). In addition to the thermogenic effect that helps to dissipate an excess of calories, brown fat is now considered to exert thermogenesis-independent effects, most likely by a host of specifically secreted factors. We will address the brown fat secretome and its specific properties in the following subsections. Initially, a brief introduction to the biology and physiology of brown fat is presented.

PHYSIOLOGY AND REGULATION OF BROWN ADIPOSE TISSUE

Our knowledge of BAT has been influenced by studies in rodent models. The murine adipose organ consists of subcutaneous and visceral depots (Cinti, 2001). In most small rodents, brown areas are recognizable in the interscapular, axillary, and cervical portions of the anterior subcutaneous depots and in the mediastinal and perirenal visceral depots. However, during the last 50 years, BAT depots were observed only in human newborns in the axillary, cervical, perirenal, and periadrenal regions (Cannon and Nedergaard, 2004) with a decrease after birth. Until 2009, there were few evidences that confirmed the existence of BAT in human muscle (Wijers et al., 2008); these data were discussed with caution and skepticism. Later, the availability of the F-fluorodeoxyglucose (F-FDG) absorption combined with the positron emission tomography/computed tomography (PET/CT) helped to define the different loci of human BAT (hBAT). This technique is currently used to detect the tumor/metastasis mass. When this method was used to detect the tumor and metastatic masses, a confounding high symmetric uptake of FDG was observed in the axillary, cervical, perirenal, and periadrenal regions. It was speculated that this might be hBAT. In 2009, five independent groups showed the presence of hBAT using the FDG-PET/CT approach, as well as the potential benefits for whole body and for energy homeostasis (Cypess et al., 2009; Saito et al., 2009; van Marken Lichtenbelt et al., 2009; Virtanen et al., 2009; Zingaretti et al., 2009). Virtanen and coworkers showed enhanced F-FDG uptake in supraclavicular area of five subjects in response to cold exposure, which was described as a putative BAT (Virtanen et al., 2009). To confirm their hypothesis, they carried out quantitative PCR analysis for a panel of genes related to BAT, such as uncoupling protein 1 (UCP1), deiodinase iodothyronine type II (DIO2), the transcriptional factors PRDM16 and PGC1α, and the β_3-adrenergic receptor (ADRB3). For all these genes they observed a significant increase in the putative regions of BAT, as compared with WAT. Western blotting and histology studies confirmed the presence of UCP1 and cytochrome c proteins in BAT selectively. Moreover, confocal microscopy analysis confirmed the colocalization of UCP1 and cytochrome oxidase subunit I (COI) proteins in mitochondrial membranes. These findings confirmed that the AT in the supraclavicular region met all criteria to be defined as BAT. The presence of BAT in supraclavicular paracervical region was also confirmed in a retrospective study (Cypess et al., 2009). The analysis of FDG-PET/CT scans of almost 2000 subjects showed an inverse correlation between BAT activity and different external cues (such as outdoor temperature, use of β-blocker drugs, and BMI), suggesting a role for SNS and for diet in the regulation of this tissue. Using similar methodologies, two other independent studies (Saito et al., 2009; van Marken Lichtenbelt et al., 2009) confirmed the inverse correlation between BAT activity and BMI, suggesting that the capacity to activate the BAT in humans is an important factor in the pathogenesis of obesity. Cinti and coworkers (Zingaretti et al., 2009) observed the presence of BAT in biopsies from perithyroid AT, detecting UCP1 in almost one-third of examined tissues.

The physiological role of BAT is related to the thermogenesis and to a broad variety of physiological conditions that modulate this process. The outstanding thermogenic potential of BAT is conferred to by UCP1 (Cannon and Nedergaard, 2004). UCP1 is a protein localized in the inner membrane of mitochondria, which uncouples mitochondrial respiration from ATP synthesis. When activated, it causes a leak that dissipates the electrochemical proton gradient that builds up across the inner mitochondrial membrane during BAT fatty acid oxidation. This gradient represents the proton-motive force used in most tissues to drive the conversion of adenosine-5'-diphosphate (ADP) to ATP by ATP synthase. Because it is found only in brown adipocytes, UCP1 constitutes the ultimate marker of these cells. The control of BAT activity depends on the adrenergic stimulation of brown adipocytes (Cannon and Nedergaard, 2004). BAT depots are densely innervated by efferent branches of the SNS, which reach individual brown adipocytes and release norepinephrine that activates UCP1.

There is controversial evidence that BAT thermogenesis participates in energy balance (Richard, 2007). From animal studies, it has been argued whether BAT thermogenesis burns off excess calories in a state of positive energy balance to maintain energy homeostasis. This concept emerged when Rothwell and Stock (1982) observed that in murine models, cafeteria diet increased both energy expenditure and BAT thermogenic activity to limit excess energetic deposition. Conversely, fat loss reduces BAT thermogenesis so that energy can be spared. A reduction in BAT thermogenesis also contributes to the positive energy balance of several obese mutants such as the leptin-deficient ob/ob mouse (Trayhurn et al., 1982) and the leptin-resistant db/db mouse. Some reports demonstrated an increased development of obesity in mice lacking BAT (Lowell et al., 1993) or UCP1 (Feldmann et al., 2009). Furthermore, UCP1-ablated mice showed a marked increase in their fat gain associated with a reduced adaptive thermogenic response of BAT only when they were housed at a temperature insuring thermoneutrality (Feldmann et al., 2009). According to these studies, there is no evidence to indicate a negative role of BAT to energy balance. Recently, a study confirmed the beneficial effects of BAT in the regulation of glucose homeostasis and insulin resistance. BAT transplantation induced positive effects on body composition, insulin sensitivity, and glucose tolerance potentially owing to increased circulating levels of IL-6 (Stanford et al., 2013).

Formation of AT starts during midgestation in higher mammals with two types of AT, WAT and BAT, originating from mesoderm. The adipogenic process has two different phases: the determination and the differentiation of stem cells. The determination of stem cells involves the commitment to the adipocyte lineage. Atit and coworkers (Atit et al., 2006) demonstrated that white and brown adipocyte lineage arise from distinct developmental programs during the embryonic development. Interestingly, microarray analysis showed that adipose stem cells from BAT express skeletal muscle–associated genes such as Myf5, MyoD, and myogenin (Timmons et al., 2007). These findings suggested that BAT and skeletal muscle share a common developmental program that is different from WAT. Studies of in vivo lineage tracing in mice confirmed that Myf5 positive cells gave rise to skeletal muscle and also to BAT (Seale et al., 2008). These results emphasize the developmental sharing among skeletal muscle and BAT and the distinct origin of WAT and BAT. The classical scenario of WAT and BAT was modified in 2010 when Petrovic et al. (2010) described a new kind of adipocytes termed as 'brite' (brown-in-white) adipocytes. These cells are ectopic brownlike cells appearing in WAT in response to cold exposure or chronic treatment with PPARγ agonists. 'Brite' adipocytes are characterized by high mitochondria number and activity, brownish color, and expression of UCP1 protein. Although they share morphological and biochemical features, as well as thermogenic potential with the classical brown adipocytes, Seale and colleagues indicated that 'brite' adipocytes do not share common origin with brown cells because their precursors do not express Myf5. Wu et al. reported that a subset of progenitor cells within WAT can differentiate in 'brite' adipocytes (Wu et al., 2012). They identified two surface proteins to select 'brite' precursors such as CD137 and TMEM26. Furthermore, these 'brite' precursor cells can fully differentiate into 'brite' adipocytes, which have the ability to switch from an energy storage to an energy dissipation state.

THE BROWN FAT SECRETOME

In contrast to WAT, which is now considered as a major endocrine organ (see earlier sections of this chapter), the potential endocrine function of BAT has been much less explored and only recent evidence supports an endocrine function of this tissue in addition to the local autocrine and paracrine functions (Villarroya et al., 2013). Strong evidence for this notion was obtained from transplantation experiments where BAT transplantation improved metabolic performance in diet-induced obese mice, an effect mediated by expression and release of IL-6 from the transplanted BAT (Stanford et al., 2013). It is therefore reasonable to believe that the brown fat secretome acts as an endocrine regulator of systemic metabolism. However, detailed studies on the brown fat secretome are still lacking (Wang et al., 2015), specifically in humans owing to the difficulties in obtaining appropriate tissue samples or cell cultures. Wang and coworkers (Wang et al., 2014) reported a distinct set of secreted factors from brown fat inducible in response to thermogenic activation. Thus, it seems unlikely that brown fat secretes highly specific adipokines, but preferred expression of certain factors may play an important role for metabolic homeostasis.

One factor that was found to be strongly induced during brown adipocyte differentiation was neuregulin 4 (Nrg4) (Wang et al., 2014). This factor could be called a "batokine" and was found to be induced during brown adipogenesis and additionally stimulated by adrenergic signaling (Rosell et al., 2014). This protein belongs to the epidermal growth factor (EGF) family of ligands and triggers downstream signaling by activation of receptor tyrosine kinases (ErbB3/4). Although it has a wide tissue distribution, its abundant expression in brown fat points to a specific function in this tissue. However, recent data show that Nrg4 is not required for cold-induced thermogenesis (Wang et al., 2014)and improves metabolic homeostasis by attenuation of hepatic lipogenesis, involving STAT5 phosphorylation and reduction of de novo lipogenesis in hepatocytes (Wang et al., 2014). Very recently it was reported that Nrg4 regulates AT secretome gene expression and adipokine secretion, promoting a healthy adipokine profile during obesity (Chen et al., 2017b). These findings are of great interest and profile Nrg4 as a positive regulator of glucose and lipid metabolism with potential impact for treating patients with NAFLD.

Another secreted protein that could play a major role in the endocrine communication from BAT to other organs is fibroblast growth factor 21 (FGF21) (Luo et al., 2017; Salminen et al., 2017a,b,c; Strowski, 2017; Maratos-Flier, 2017; Sonoda et al., 2017; Salminen et al., 2017c). This protein is preferentially expressed in the liver and may also play a role as a myokine; however, it was shown that BAT also contributes to plasma FGF21 on cold exposure. It could be demonstrated that FGF21 from activated BAT can enter the circulation and, at least partly, mediates the rise of plasma FGF21 after cold exposure (Hondares et al., 2011), both in rodents and humans. Solid evidence exists that FGF21 exerts a positive effect on glucose and lipid metabolism, specifically involving hepatic β-oxidation and gluconeogenesis (Badman et al., 2007). Furthermore, application of FGF21 was found to correct obesity and improve metabolic homeostasis in humans (Gaich et al., 2013). FGF21 also acts on AT in an autocrine/paracrine fashion and was shown to stimulate the thermogenic program in brown fat cells and to induce white fat browning (Fisher et al., 2012).

The so-called BMPs play a key role in the regulation of bone and cartilage formation and repair. These proteins belong to the TGF-β superfamily and have been recognized to be also relevant for the regulation of adipocyte differentiation and energy expenditure (Zamani and Brown, 2011). BMP7 was described as a regulator of mitochondrial biogenesis during brown adipocyte differentiation; furthermore, BMP7 was also found to promote the browning of white fat (Schulz et al., 2011). In humans, BMP4 was also able to promote the browning of white adipocytes (Elsen et al., 2014), whereas this effect was not observed in rodents.

A very intriguing example of a protein exerting pleiotropic and multiple crosstalk functions is the cytokine IL-6. IL-6 is expressed and released by a number of tissues including WAT (see earlier sections of this chapter), skeletal muscle (see Chapter 3), and immune cells. In addition to being a mediator of inflammation, IL-6 also functions as a major regulator of glucose metabolism and energy homeostasis (Pal et al., 2014). The knockout of IL-6 leads to glucose intolerance under high-fat diet, and IL-6 is known to be regulated by exercise, representing the paradigm of a myokine (see Chapter 3). In fact, IL-6 must be considered as the prototype of an adipomyokine, and this specific aspect will be further discussed in Chapter 4. Interestingly, BAT-derived IL-6 appears to be the mediator of the improvement of metabolic performance after brown fat transplantation (Stanford et al., 2013). This issue needs further investigation and may provide an even more complex view on the biology of IL-6.

REFERENCES

Abraham, N.G., Kappas, A., 2008. Pharmacological and clinical aspects of heme oxygenase. Pharmacol. Rev. 60, 79–127.

Ahima, R.S., Flier, J.S., 2000. Leptin. Annu. Rev. Physiol. 62, 413–437.

Ahren, B., Larsson, H., 1997. Leptin–a regulator of islet function: its plasma levels correlate with glucagon and insulin secretion in healthy women. Metabolism 46, 1477–1481.

Aldiss, P., Dellschaft, N., Sacks, H., Budge, H., Symonds, M.E., 2017. Beyond obesity - thermogenic adipocytes and cardiometabolic health. Horm. Mol. Biol. Clin. Investig. 31.

Ali, A.T., Hochfeld, W.E., Myburgh, R., Pepper, M.S., 2013. Adipocyte and adipogenesis. Eur. J. Cell Biol. 92, 229–236.

Arita, Y., Kihara, S., Ouchi, N., Maeda, K., Kuriyama, H., Okamoto, Y., Kumada, M., Hotta, K., Nishida, M., Takahashi, M., Nakamura, T., Shimomura, I., Muraguchi, M., Ohmoto, Y., Funahashi, T., Matsuzawa, Y., 2002. Adipocyte-derived plasma protein adiponectin acts as a platelet-derived growth factor-BB-binding protein and regulates growth factor-induced common postreceptor signal in vascular smooth muscle cell. Circulation 105, 2893–2898.

Arounleut, P., Bowser, M., Upadhyay, S., Shi, X.M., Fulzele, S., Johnson, M.H., Stranahan, A.M., Hill, W.D., Isales, C.M., Hamrick, M.W., 2013. Absence of functional leptin receptor isoforms in the POUND (Lepr(db/lb)) mouse is associated with muscle atrophy and altered myoblast proliferation and differentiation. PLoS One 8, e72330.

Atit, R., Sgaier, S.K., Mohamed, O.A., Taketo, M.M., Dufort, D., Joyner, A.L., Niswander, L., Conlon, R.A., 2006. Beta-catenin activation is necessary and sufficient to specify the dorsal dermal fate in the mouse. Dev. Biol. 296, 164–176.

Ayala-Lopez, N., Watts, S.W., 2017. New actions of an old friend: perivascular adipose tissue's adrenergic mechanisms. Br. J. Pharmacol. 174, 3454–3465.

Badman, M.K., Pissios, P., Kennedy, A.R., Koukos, G., Flier, J.S., Maratos-Flier, E., 2007. Hepatic fibroblast growth factor 21 is regulated by PPARalpha and is a key mediator of hepatic lipid metabolism in ketotic states. Cell Metab. 5, 426–437.

Bao, W., Song, F., Li, X., Rong, S., Yang, W., Zhang, M., Yao, P., Hao, L., Yang, N., Hu, F.B., Liu, L., 2010. Plasma heme oxygenase-1 concentration is elevated in individuals with type 2 diabetes mellitus. PLoS One 5, e12371.

Barber, M.C., Ward, R.J., Richards, S.E., Salter, A.M., Buttery, P.J., Vernon, R.G., Travers, M.T., 2000. Ovine adipose tissue monounsaturated fat content is correlated to depot-specific expression of the stearoyl-CoA desaturase gene. J. Anim. Sci. 78, 62–68.

Bauer, S., Wanninger, J., Schmidhofer, S., Weigert, J., Neumeier, M., Dorn, C., Hellerbrand, C., Zimara, N., Schaffler, A., Aslanidis, C., Buechler, C., 2011. Sterol regulatory element-binding protein 2 (SREBP2) activation after excess triglyceride storage induces chemerin in hypertrophic adipocytes. Endocrinology 152, 26–35.

Bays, H.E., Gonzalez-Campoy, J.M., Bray, G.A., Kitabchi, A.E., Bergman, D.A., Schorr, A.B., Rodbard, H.W., Henry, R.R., 2008. Pathogenic potential of adipose tissue and metabolic consequences of adipocyte hypertrophy and increased visceral adiposity. Expert Rev. Cardiovasc Ther. 6, 343—368.

Beltowski, J., 2012. Leptin and the regulation of endothelial function in physiological and pathological conditions. Clin. Exp. Pharmacol. Physiol. 39, 168—178.

Berry, R., Jeffery, E., Rodeheffer, M.S., 2014. Weighing in on adipocyte precursors. Cell Metab. 19, 8—20.

Berry, R., Rodeheffer, M.S., 2013. Characterization of the adipocyte cellular lineage in vivo. Nat. Cell Biol. 15, 302—308.

Betz, M.J., Enerback, S., 2017. Targeting thermogenesis in brown fat and muscle to treat obesity and metabolic disease. Nat. Rev. Endocrinol.

Black, R.A., Rauch, C.T., Kozlosky, C.J., Peschon, J.J., Slack, J.L., Wolfson, M.F., Castner, B.J., Stocking, K.L., Reddy, P., Srinivasan, S., Nelson, N., Boiani, N., Schooley, K.A., Gerhart, M., Davis, R., Fitzner, J.N., Johnson, R.S., Paxton, R.J., March, C.J., Cerretti, D.P., 1997. A metalloproteinase disintegrin that releases tumour-necrosis factor-alpha from cells. Nature 385, 729—733.

Bluher, M., 2010. The distinction of metabolically 'healthy' from 'unhealthy' obese individuals. Curr. Opin. Lipidol. 21, 38—43.

Blüher, M., 2012. Are there still healthy obese patients? Curr. Opin. Endocrinol. Diabetes Obes. 19, 341—346.

Blüher, M., 2014. Are metabolically healthy obese individuals really healthy? Eur. J. Endocrinol. 171, R209—R219.

Bondue, B., Wittamer, V., Parmentier, M., 2011. Chemerin and its receptors in leukocyte trafficking, inflammation and metabolism. Cytokine Growth Factor Rev. 22, 331—338.

Bose, A.K., Mocanu, M.M., Carr, R.D., Brand, C.L., Yellon, D.M., 2005. Glucagon-like peptide 1 can directly protect the heart against ischemia/reperfusion injury. Diabetes 54, 146—151.

Brough, D., Rothwell, N.J., 2007. Caspase-1-dependent processing of pro-interleukin-1beta is cytosolic and precedes cell death. J. Cell Sci. 120, 772—781.

Brown, J.E., Dunmore, S.J., 2007. Leptin decreases apoptosis and alters BCL-2 : bax ratio in clonal rodent pancreatic beta-cells. Diabetes Metab. Res. Rev. 23, 497—502.

Buscemi, S., Corleo, D., Buscemi, C., Giordano, C., 2017. Does iris(in) bring bad news or good news? Eat Weight Disord.

Cammisotto, P.G., Gelinas, Y., Deshaies, Y., Bukowiecki, L.J., 2005. Regulation of leptin secretion from white adipocytes by insulin, glycolytic substrates, and amino acids. Am. J. Physiol. Endocrinol. Metab. 289, E166—E171.

Cannon, B., Nedergaard, J., 2004. Brown adipose tissue: function and physiological significance. Physiol. Rev. 84, 277—359.

Carobbio, S., Pellegrinelli, V., Vidal-Puig, A., 2017. Adipose tissue function and expandability as determinants of lipotoxicity and the metabolic syndrome. Adv. Exp. Med. Biol. 960, 161—196.

Carson, B.P., Del Bas, J.M., Moreno-Navarrete, J.M., Fernandez-Real, J.M., Mora, S., 2013. The rab11 effector protein FIP1 regulates adiponectin trafficking and secretion. PLoS One 8, e74687.

Carta, S., Lavieri, R., Rubartelli, A., 2013. Different members of the IL-1 family come out in different ways: DAMPs vs. Cytokines? Front. Immunol. 4, 123.

Catoire, M., Kersten, S., 2015. The search for exercise factors in humans. Faseb J. 29, 1615−1628.

Caton, P.W., Kieswich, J., Yaqoob, M.M., Holness, M.J., Sugden, M.C., 2011. Nicotinamide mononucleotide protects against pro-inflammatory cytokine-mediated impairment of mouse islet function. Diabetologia 54, 3083−3092.

Cawthorn, W.P., Sethi, J.K., 2008. TNF-alpha and adipocyte biology. FEBS Lett. 582, 117−131.

Cernea, S., Raz, I., 2011. Therapy in the early stage: incretins. Diabetes Care 34 (Suppl. 2), S264−S271.

Chatterjee, T.K., Aronow, B.J., Tong, W.S., Manka, D., Tang, Y., Bogdanov, V.Y., Unruh, D., Blomkalns, A.L., Piegore Jr., M.G., Weintraub, D.S., Rudich, S.M., Kuhel, D.G., Hui, D.Y., Weintraub, N.L., 2013. Human coronary artery perivascular adipocytes overexpress genes responsible for regulating vascular morphology, inflammation, and hemostasis. Physiol. Genomics 45, 697−709.

Chen, Y., Pan, R., Pfeifer, A., 2016. Fat tissues, the brite and the dark sides. Pflugers Arch. 468, 1803−1807.

Chen, Y., Pan, R., Pfeifer, A., 2017a. Regulation of brown and beige fat by microRNAs. Pharmacol. Ther. 170, 1−7.

Chen, Y.W., Gregory, C.M., Scarborough, M.T., Shi, R., Walter, G.A., Vandenborne, K., 2007. Transcriptional pathways associated with skeletal muscle disuse atrophy in humans. Physiol. Genomics 31, 510−520.

Chen, Z., Wang, G.X., Ma, S.L., Jung, D.Y., Ha, H., Altamimi, T., Zhao, X.Y., Guo, L., Zhang, P., Hu, C.R., Cheng, J.X., Lopaschuk, G.D., Kim, J.K., Lin, J.D., 2017b. Nrg4 promotes fuel oxidation and a healthy adipokine profile to ameliorate diet-induced metabolic disorders. Mol. Metab. 6, 863−872.

Cheng, J., Song, Z.Y., Pu, L., Yang, H., Zheng, J.M., Zhang, Z.Y., Shi, X.E., Yang, G.S., 2013. Retinol binding protein 4 affects the adipogenesis of porcine preadipocytes through insulin signaling pathways. Biochem. Cell Biol. 91, 236−243.

Cheng, Q., Dong, W., Qian, L., Wu, J., Peng, Y., 2011. Visfatin inhibits apoptosis of pancreatic beta-cell line, Min6, via the mitogen-activated protein kinase/phosphoinositide 3-kinase pathway. J. Mol. Endocrinol. 47, 13−21.

Chistiakov, D.A., Grechko, A.V., Myasoedova, V.A., Melnichenko, A.A., Orekhov, A.N., 2017. Impact of the cardiovascular system-associated adipose tissue on atherosclerotic pathology. Atherosclerosis 263, 361−368.

Choe, S.S., Huh, J.Y., Hwang, I.J., Kim, J.I., Kim, J.B., 2016. Adipose tissue remodeling: its role in energy metabolism and metabolic disorders. Front. Endocrinol. (Lausanne) 7, 30.

Chu, D.T., Gawronska-Kozak, B., 2017. Brown and brite adipocytes: same function, but different origin and response. Biochimie 138, 102−105.

Chu, D.T., Tao, Y., 2017. Human thermogenic adipocytes: a reflection on types of adipocyte, developmental origin, and potential application. J. Physiol. Biochem. 73, 1−4.

Cinti, S., 2001. The adipose organ: morphological perspectives of adipose tissues. Proc. Nutr. Soc. 60, 319−328.

Cinti, S., 2009. Transdifferentiation properties of adipocytes in the adipose organ. Am. J. Physiol. Endocrinol. Metab. 297, E977−E986.

Cinti, S., 2017. UCP1 protein: the molecular hub of adipose organ plasticity. Biochimie 134, 71−76.

Cohen, P., Spiegelman, B.M., 2016. Cell biology of fat storage. Mol. Biol. Cell 27, 2523–2527.

Cohen, P., Yang, G., Yu, X., Soukas, A.A., Wolfish, C.S., Friedman, J.M., Li, C., 2005. Induction of leptin receptor expression in the liver by leptin and food deprivation. J. Biol. Chem. 280, 10034–10039.

Coleman, D.L., 1982. Diabetes-obesity syndromes in mice. Diabetes 31, 1–6.

Contreras, C., Nogueiras, R., Dieguez, C., Rahmouni, K., Lopez, M., 2017a. Traveling from the hypothalamus to the adipose tissue: the thermogenic pathway. Redox Biol. 12, 854–863.

Contreras, G.A., Strieder-Barboza, C., Raphael, W., 2017b. Adipose tissue lipolysis and remodeling during the transition period of dairy cows. J. Anim. Sci. Biotechnol. 8, 41.

Costford, S.R., Bajpeyi, S., Pasarica, M., Albarado, D.C., Thomas, S.C., Xie, H., Church, T.S., Jubrias, S.A., Conley, K.E., Smith, S.R., 2010. 'Skeletal muscle NAMPT is induced by exercise in humans. Am. J. Physiol. Endocrinol. Metab. 298, E117–E126.

Crewe, C., An, Y.A., Scherer, P.E., 2017. The ominous triad of adipose tissue dysfunction: inflammation, fibrosis, and impaired angiogenesis. J. Clin. Invest. 127, 74–82.

Cusi, K., 2009. Nonalcoholic fatty liver disease in type 2 diabetes mellitus. Curr. Opin. Endocrinol. Diabetes Obes. 16, 141–149.

Cusi, K., 2010. The role of adipose tissue and lipotoxicity in the pathogenesis of type 2 diabetes. Curr. Diab Rep. 10, 306–315.

Cypess, A.M., Lehman, S., Williams, G., Tal, I., Rodman, D., Goldfine, A.B., Kuo, F.C., Palmer, E.L., Tseng, Y.H., Doria, A., Kolodny, G.M., Kahn, C.R., 2009. Identification and importance of brown adipose tissue in adult humans. N. Engl. J. Med. 360, 1509–1517.

Dahl, J.P., Binda, A., Canfield, V.A., Levenson, R., 2000. Participation of Na,K-ATPase in FGF-2 secretion: rescue of ouabain-inhibitable FGF-2 secretion by ouabain-resistant Na,K-ATPase alpha subunits. Biochemistry 39, 14877–14883.

Dahlman, I., Elsen, M., Tennagels, N., Korn, M., Brockmann, B., Sell, H., Eckel, J., Arner, P., 2012. Functional annotation of the human fat cell secretome. Arch. Physiol. Biochem. 118, 84–91.

Dattaroy, D., Pourhoseini, S., Das, S., Alhasson, F., Seth, R.K., Nagarkatti, M., Michelotti, G.A., Diehl, A.M., Chatterjee, S., 2015. Micro-RNA 21 inhibition of SMAD7 enhances fibrogenesis via leptin-mediated NADPH oxidase in experimental and human nonalcoholic steatohepatitis. Am. J. Physiol. Gastrointest. Liver Physiol. 308, G298–G312.

Deshmukh, A.S., 2016. Insulin-stimulated glucose uptake in healthy and insulin-resistant skeletal muscle. Horm. Mol. Biol. Clin. Investig. 26, 13–24.

Dietze, D., Koenen, M., Rohrig, K., Horikoshi, H., Hauner, H., Eckel, J., 2002. Impairment of insulin signaling in human skeletal muscle cells by co-culture with human adipocytes. Diabetes 51, 2369–2376.

Dietze-Schroeder, D., Sell, H., Uhlig, M., Koenen, M., Eckel, J., 2005. Autocrine action of adiponectin on human fat cells prevents the release of insulin resistance-inducing factors. Diabetes 54, 2003–2011.

Dinarello, C.A., 1987. Interleukin-1: multiple biological properties and mechanisms of action. Adv. Prostaglandin Thromboxane Leukot. Res. 17B, 900–904.

do Carmo Avides, M., Domingues, L., Vicente, A., Teixeira, J., 2008. Differentiation of human pre-adipocytes by recombinant adiponectin. Protein Expr. Purif. 59, 122–126.

Drucker, D.J., Nauck, M.A., 2006. The incretin system: glucagon-like peptide-1 receptor agonists and dipeptidyl peptidase-4 inhibitors in type 2 diabetes. Lancet 368, 1696−1705.

Dubrovska, G., Verlohren, S., Luft, F.C., Gollasch, M., 2004. Mechanisms of ADRF release from rat aortic adventitial adipose tissue. Am. J. Physiol. Heart Circ. Physiol. 286, H1107−H1113.

Dunbar, J.C., Walsh, M.F., 1980. Glucagon and insulin secretion by islets of lean and obese (ob/ob) mice. Horm. Metab. Res. 12, 39−40.

Dunmore, S.J., Brown, J.E., 2013. The role of adipokines in beta-cell failure of type 2 diabetes. J. Endocrinol. 216, T37−T45.

Dutour, A., Achard, V., Sell, H., Naour, N., Collart, F., Gaborit, B., Silaghi, A., Eckel, J., Alessi, M.C., Henegar, C., Clement, K., 2010. Secretory type II phospholipase A2 is produced and secreted by epicardial adipose tissue and overexpressed in patients with coronary artery disease. J. Clin. Endocrinol. Metab. 95, 963−967.

Eder, C., 2009. Mechanisms of interleukin-1beta release. Immunobiology 214, 543−553.

Ehrlund, A., Mejhert, N., Lorente-Cebrian, S., Astrom, G., Dahlman, I., Laurencikiene, J., Ryden, M., 2013. Characterization of the Wnt inhibitors secreted frizzled-related proteins (SFRPs) in human adipose tissue. J. Clin. Endocrinol. Metab. 98, E503−E508.

El Hadi, H., Vettor, R., Rossato, M., 2017. Functional imaging of brown adipose tissue in human. Horm. Mol. Biol. Clin. Investig. 31.

Elsen, M., Raschke, S., Tennagels, N., Schwahn, U., Jelenik, T., Roden, M., Romacho, T., Eckel, J., 2014. BMP4 and BMP7 induce the white-to-brown transition of primary human adipose stem cells. Am. J. Physiol. Cell Physiol. 306, C431−C440.

Fadini, G.P., Avogaro, A., 2013. Dipeptidyl peptidase-4 inhibition and vascular repair by mobilization of endogenous stem cells in diabetes and beyond. Atherosclerosis 229, 23−29.

Fain, J.N., Sacks, H.S., Bahouth, S.W., Tichansky, D.S., Madan, A.K., Cheema, P.S., 2010. Human epicardial adipokine messenger RNAs: comparisons of their expression in substernal, subcutaneous, and omental fat. Metabolism 59, 1379−1386.

Fasshauer, M., Bluher, M., 2015. Adipokines in health and disease. Trends Pharmacol. Sci. 36, 461−470.

Feldmann, H.M., Golozoubova, V., Cannon, B., Nedergaard, J., 2009. UCP1 ablation induces obesity and abolishes diet-induced thermogenesis in mice exempt from thermal stress by living at thermoneutrality. Cell Metab. 9, 203−209.

Fernandez-Quintela, A., Milton-Laskibar, I., Gonzalez, M., Portillo, M.P., 2017. Antiobesity effects of resveratrol: which tissues are involved? Ann. N. Y. Acad. Sci. 1403, 118−131.

Fesus, G., Dubrovska, G., Gorzelniak, K., Kluge, R., Huang, Y., Luft, F.C., Gollasch, M., 2007. Adiponectin is a novel humoral vasodilator. Cardiovasc Res. 75, 719−727.

Fisher, F.M., Kleiner, S., Douris, N., Fox, E.C., Mepani, R.J., Verdeguer, F., Wu, J., Kharitonenkov, A., Flier, J.S., Maratos-Flier, E., Spiegelman, B.M., 2012. FGF21 regulates PGC-1alpha and browning of white adipose tissues in adaptive thermogenesis. Genes Dev. 26, 271−281.

Flak, J.N., Myers Jr., M.G., 2016. Minireview: CNS mechanisms of leptin action. Mol. Endocrinol. 30, 3−12.

Florkiewicz, R.Z., Majack, R.A., Buechler, R.D., Florkiewicz, E., 1995. Quantitative export of FGF-2 occurs through an alternative, energy-dependent, non-ER/Golgi pathway. J. Cell Physiol. 162, 388−399.

Flouris, A.D., Dinas, P.C., Valente, A., Andrade, C.M.B., Kawashita, N.H., Sakellariou, P., 2017. Exercise-induced effects on UCP1 expression in classical brown adipose tissue: a systematic review. Horm. Mol. Biol. Clin. Investig. 31.

Fu, Y., Luo, N., Klein, R.L., Garvey, W.T., 2005. Adiponectin promotes adipocyte differentiation, insulin sensitivity, and lipid accumulation. J. Lipid Res. 46, 1369—1379.

Fukuhara, A., Matsuda, M., Nishizawa, M., Segawa, K., Tanaka, M., Kishimoto, K., Matsuki, Y., Murakami, M., Ichisaka, T., Murakami, H., Watanabe, E., Takagi, T., Akiyoshi, M., Ohtsubo, T., Kihara, S., Yamashita, S., Makishima, M., Funahashi, T., Yamanaka, S., Hiramatsu, R., Matsuzawa, Y., Shimomura, I., 2007. Retraction. Science 318, 565.

Gaich, G., Chien, J.Y., Fu, H., Glass, L.C., Deeg, M.A., Holland, W.L., Kharitonenkov, A., Bumol, T., Schilske, H.K., Moller, D.E., 2013. The effects of LY2405319, an FGF21 analog, in obese human subjects with type 2 diabetes. Cell Metab. 18, 333—340.

Garber, A.J., 2011. Incretin effects on beta-cell function, replication, and mass: the human perspective. Diabetes Care 34 (Suppl. 2), S258—S263.

Giannocco, G., Oliveira, K.C., Crajoinas, R.O., Venturini, G., Salles, T.A., Fonseca-Alaniz, M.H., Maciel, R.M., Girardi, A.C., 2013. Dipeptidyl peptidase IV inhibition upregulates GLUT4 translocation and expression in heart and skeletal muscle of spontaneously hypertensive rats. Eur. J. Pharmacol. 698, 74—86.

Gil-Ortega, M., Somoza, B., Huang, Y., Gollasch, M., Fernandez-Alfonso, M.S., 2015. Regional differences in perivascular adipose tissue impacting vascular homeostasis. Trends Endocrinol. Metab. 26, 367—375.

Gollasch, M., 2012. Vasodilator signals from perivascular adipose tissue. Br. J. Pharmacol. 165, 633—642.

Gollasch, M., Dubrovska, G., 2004. Paracrine role for periadventitial adipose tissue in the regulation of arterial tone. Trends Pharmacol. Sci. 25, 647—653.

Gomez-Navarro, N., Miller, E., 2016. Protein sorting at the ER-Golgi interface. J. Cell Biol. 215, 769—778.

Goralski, K.B., McCarthy, T.C., Hanniman, E.A., Zabel, B.A., Butcher, E.C., Parlee, S.D., Muruganandan, S., Sinal, C.J., 2007. Chemerin, a novel adipokine that regulates adipogenesis and adipocyte metabolism. J. Biol. Chem. 282, 28175—28188.

Granneman, J.G., Li, P., Zhu, Z., Lu, Y., 2005. Metabolic and cellular plasticity in white adipose tissue I: effects of beta3-adrenergic receptor activation. Am. J. Physiol. Endocrinol. Metab. 289, E608—E616.

Greulich, S., Chen, W.J., Maxhera, B., Rijzewijk, L.J., van der Meer, R.W., Jonker, J.T., Mueller, H., de Wiza, D.H., Floerke, R.R., Smiris, K., Lamb, H.J., de Roos, A., Bax, J.J., Romijn, J.A., Smit, J.W., Akhyari, P., Lichtenberg, A., Eckel, J., Diamant, M., Ouwens, D.M., 2013. Cardioprotective properties of omentin-1 in type 2 diabetes: evidence from clinical and in vitro studies. PLoS One 8, e59697.

Greulich, S., de Wiza, D.H., Preilowski, S., Ding, Z., Mueller, H., Langin, D., Jaquet, K., Ouwens, D.M., Eckel, J., 2011. Secretory products of Guinea pig epicardial fat induce insulin resistance and impair primary adult rat cardiomyocyte function. J. Cell Mol. Med. 15, 2399—2410.

Greulich, S., Maxhera, B., Vandenplas, G., de Wiza, D.H., Smiris, K., Mueller, H., Heinrichs, J., Blumensatt, M., Cuvelier, C., Akhyari, P., Ruige, J.B., Ouwens, D.M., Eckel, J., 2012. Secretory products from epicardial adipose tissue of patients with type 2 diabetes mellitus induce cardiomyocyte dysfunction. Circulation 126, 2324—2334.

Gunnett, C.A., Heistad, D.D., Faraci, F.M., 2003. Gene-targeted mice reveal a critical role for inducible nitric oxide synthase in vascular dysfunction during diabetes. Stroke 34, 2970—2974.

Guzik, T.J., Mangalat, D., Korbut, R., 2006. Adipocytokines - novel link between inflammation and vascular function? J. Physiol. Pharmacol. 57, 505—528.

Hammarstedt, A., Hedjazifar, S., Jenndahl, L., Gogg, S., Grunberg, J., Gustafson, B., Klimcakova, E., Stich, V., Langin, D., Laakso, M., Smith, U., 2013. WISP2 regulates preadipocyte commitment and PPARgamma activation by BMP4. Proc. Natl. Acad. Sci. U. S. A. 110, 2563−2568.

Heiker, J.T., Kloting, N., Kovacs, P., Kuettner, E.B., Strater, N., Schultz, S., Kern, M., Stumvoll, M., Bluher, M., Beck-Sickinger, A.G., 2013. Vaspin inhibits kallikrein 7 by serpin mechanism. Cell Mol. Life Sci. 70, 2569−2583.

Hida, K., Wada, J., Eguchi, J., Zhang, H., Baba, M., Seida, A., Hashimoto, I., Okada, T., Yasuhara, A., Nakatsuka, A., Shikata, K., Hourai, S., Futami, J., Watanabe, E., Matsuki, Y., Hiramatsu, R., Akagi, S., Makino, H., Kanwar, Y.S., 2005. Visceral adipose tissue-derived serine protease inhibitor: a unique insulin-sensitizing adipocytokine in obesity. Proc. Natl. Acad. Sci. U. S. A. 102, 10610−10615.

Holland, W.L., Miller, R.A., Wang, Z.V., Sun, K., Barth, B.M., Bui, H.H., Davis, K.E., Bikman, B.T., Halberg, N., Rutkowski, J.M., Wade, M.R., Tenorio, V.M., Kuo, M.S., Brozinick, J.T., Zhang, B.B., Birnbaum, M.J., Summers, S.A., Scherer, P.E., 2011. Receptor-mediated activation of ceramidase activity initiates the pleiotropic actions of adiponectin. Nat. Med. 17, 55−63.

Hondares, E., Iglesias, R., Giralt, A., Gonzalez, F.J., Giralt, M., Mampel, T., Villarroya, F., 2011. Thermogenic activation induces FGF21 expression and release in brown adipose tissue. J. Biol. Chem. 286, 12983−12990.

Hotamisligil, G.S., Budavari, A., Murray, D., Spiegelman, B.M., 1994a. Reduced tyrosine kinase activity of the insulin receptor in obesity-diabetes. Central role of tumor necrosis factor-alpha. J. Clin. Invest. 94, 1543−1549.

Hotamisligil, G.S., Murray, D.L., Choy, L.N., Spiegelman, B.M., 1994b. Tumor necrosis factor alpha inhibits signaling from the insulin receptor. Proc. Natl. Acad. Sci. U. S. A. 91, 4854−4858.

Hotamisligil, G.S., Peraldi, P., Budavari, A., Ellis, R., White, M.F., Spiegelman, B.M., 1996. IRS-1-mediated inhibition of insulin receptor tyrosine kinase activity in TNF-alpha- and obesity-induced insulin resistance. Science 271, 665−668.

Hotamisligil, G.S., Shargill, N.S., Spiegelman, B.M., 1993. Adipose expression of tumor necrosis factor-alpha: direct role in obesity-linked insulin resistance. Science 259, 87−91.

Hotamisligil, G.S., Spiegelman, B.M., 1994. Tumor necrosis factor alpha: a key component of the obesity-diabetes link. Diabetes 43, 1271−1278.

Huang, J.Y., Chiang, M.T., Chau, L.Y., 2013. Adipose overexpression of heme oxygenase-1 does not protect against high fat diet-induced insulin resistance in mice. PLoS One 8, e55369.

Huang, S., Czech, M.P., 2007. The GLUT4 glucose transporter. Cell Metab. 5, 237−252.

Huang, W., Dedousis, N., Bandi, A., Lopaschuk, G.D., O'Doherty, R.M., 2006. Liver triglyceride secretion and lipid oxidative metabolism are rapidly altered by leptin in vivo. Endocrinology 147, 1480−1487.

Huang-Doran, I., Zhang, C.Y., Vidal-Puig, A., 2017. Extracellular vesicles: novel mediators of cell communication in metabolic disease. Trends Endocrinol. Metab. 28, 3−18.

Hutley, L.J., Newell, F.S., Kim, Y.H., Luo, X., Widberg, C.H., Shurety, W., Prins, J.B., Whitehead, J.P., 2011. A putative role for endogenous FGF-2 in FGF-1 mediated differentiation of human preadipocytes. Mol. Cell Endocrinol. 339, 165−171.

Huynh, F.K., Neumann, U.H., Wang, Y., Rodrigues, B., Kieffer, T.J., Covey, S.D., 2013. A role for hepatic leptin signaling in lipid metabolism via altered very low density lipoprotein composition and liver lipase activity in mice. Hepatology 57, 543−554.

Iacobellis, G., 2016. Epicardial fat: a new cardiovascular therapeutic target. Curr. Opin. Pharmacol. 27, 13—18.

Ishibashi, Y., Matsui, T., Maeda, S., Higashimoto, Y., Yamagishi, S., 2013. Advanced glycation end products evoke endothelial cell damage by stimulating soluble dipeptidyl peptidase-4 production and its interaction with mannose 6-phosphate/insulin-like growth factor II receptor. Cardiovasc Diabetol. 12, 125.

Iwabu, M., Yamauchi, T., Okada-Iwabu, M., Sato, K., Nakagawa, T., Funata, M., Yamaguchi, M., Namiki, S., Nakayama, R., Tabata, M., Ogata, H., Kubota, N., Takamoto, I., Hayashi, Y.K., Yamauchi, N., Waki, H., Fukayama, M., Nishino, I., Tokuyama, K., Ueki, K., Oike, Y., Ishii, S., Hirose, K., Shimizu, T., Touhara, K., Kadowaki, T., 2010. Adiponectin and AdipoR1 regulate PGC-1alpha and mitochondria by Ca(2+) and AMPK/SIRT1. Nature 464, 1313—1319.

Iwaki-Egawa, S., Watanabe, Y., Kikuya, Y., Fujimoto, Y., 1998. Dipeptidyl peptidase IV from human serum: purification, characterization, and N-terminal amino acid sequence. J. Biochem. 124, 428—433.

Jackson, A., Vayssiere, B., Garcia, T., Newell, W., Baron, R., Roman-Roman, S., Rawadi, G., 2005. Gene array analysis of Wnt-regulated genes in C3H10T1/2 cells. Bone 36, 585—598.

Kakudo, N., Shimotsuma, A., Kusumoto, K., 2007. Fibroblast growth factor-2 stimulates adipogenic differentiation of human adipose-derived stem cells. Biochem. Biophys. Res. Commun. 359, 239—244.

Kiefer, F.W., 2017. The significance of beige and brown fat in humans. Endocr. Connect. 6, R70—R79.

Kim, D.H., Burgess, A.P., Li, M., Tsenovoy, P.L., Addabbo, F., McClung, J.A., Puri, N., Abraham, N.G., 2008. Heme oxygenase-mediated increases in adiponectin decrease fat content and inflammatory cytokines tumor necrosis factor-alpha and interleukin-6 in Zucker rats and reduce adipogenesis in human mesenchymal stem cells. J. Pharmacol. Exp. Ther. 325, 833—840.

Knudson, J.D., Dincer, U.D., Zhang, C., Swafford Jr., A.N., Koshida, R., Picchi, A., Focardi, M., Dick, G.M., Tune, J.D., 2005. Leptin receptors are expressed in coronary arteries, and hyperleptinemia causes significant coronary endothelial dysfunction. Am. J. Physiol. Heart Circ. Physiol. 289, H48—H56.

Kotnik, P., Keuper, M., Wabitsch, M., Fischer-Posovszky, P., 2013. Interleukin-1beta downregulates RBP4 secretion in human adipocytes. PLoS One 8, e57796.

Krijnen, P.A., Hahn, N.E., Kholova, I., Baylan, U., Sipkens, J.A., van Alphen, F.P., Vonk, A.B., Simsek, S., Meischl, C., Schalkwijk, C.G., van Buul, J.D., van Hinsbergh, V.W., Niessen, H.W., 2012. Loss of DPP4 activity is related to a prothrombogenic status of endothelial cells: implications for the coronary microvasculature of myocardial infarction patients. Basic Res. Cardiol. 107, 233.

Krook, A., Bjornholm, M., Galuska, D., Jiang, X.J., Fahlman, R., Myers Jr., M.G., Wallberg-Henriksson, H., Zierath, J.R., 2000. Characterization of signal transduction and glucose transport in skeletal muscle from type 2 diabetic patients. Diabetes 49, 284—292.

Kumada, M., Kihara, S., Sumitsuji, S., Kawamoto, T., Matsumoto, S., Ouchi, N., Arita, Y., Okamoto, Y., Shimomura, I., Hiraoka, H., Nakamura, T., Funahashi, T., Matsuzawa, Y., Study Group Coronary Artery Disease Osaka, C.A.D., 2003. Association of hypoadiponectinemia with coronary artery disease in men. Arterioscler. Thromb. Vasc. Biol. 23, 85—89.

Lagathu, C., Christodoulides, C., Tan, C.Y., Virtue, S., Laudes, M., Campbell, M., Ishikawa, K., Ortega, F., Tinahones, F.J., Fernandez-Real, J.M., Oresic, M., Sethi, J.K., Vidal-Puig, A., 2010. Secreted frizzled-related protein 1 regulates adipose tissue expansion and is dysregulated in severe obesity. Int. J. Obes. (Lond) 34, 1695−1705.

Lamers, D., Famulla, S., Wronkowitz, N., Hartwig, S., Lehr, S., Ouwens, D.M., Eckardt, K., Kaufman, J.M., Ryden, M., Muller, S., Hanisch, F.G., Ruige, J., Arner, P., Sell, H., Eckel, J., 2011. Dipeptidyl peptidase 4 is a novel adipokine potentially linking obesity to the metabolic syndrome. Diabetes 60, 1917−1925.

Lee, D.E., Kehlenbrink, S., Lee, H., Hawkins, M., Yudkin, J.S., 2009. Getting the message across: mechanisms of physiological cross talk by adipose tissue. Am. J. Physiol. Endocrinol. Metab. 296, E1210−E1229.

Lefterova, M.I., Lazar, M.A., 2009. New developments in adipogenesis. Trends Endocrinol. Metab. 20, 107−114.

Lehr, S., Hartwig, S., Lamers, D., Famulla, S., Muller, S., Hanisch, F.G., Cuvelier, C., Ruige, J., Eckardt, K., Ouwens, D.M., Sell, H., Eckel, J., 2012. Identification and validation of novel adipokines released from primary human adipocytes. Mol. Cell Proteomics 11. M111 010504.

Li, F., Li, Y., Duan, Y., Hu, C.A., Tang, Y., Yin, Y., 2017. Myokines and adipokines: involvement in the crosstalk between skeletal muscle and adipose tissue. Cytokine Growth Factor Rev. 33, 73−82.

Liu, Y., Chewchuk, S., Lavigne, C., Brule, S., Pilon, G., Houde, V., Xu, A., Marette, A., Sweeney, G., 2009. Functional significance of skeletal muscle adiponectin production, changes in animal models of obesity and diabetes, and regulation by rosiglitazone treatment. Am. J. Physiol. Endocrinol. Metab. 297, E657−E664.

Lopaschuk, G.D., 2002. Metabolic abnormalities in the diabetic heart. Heart Fail Rev. 7, 149−159.

Lopez-Bermejo, A., Chico-Julia, B., Fernandez-Balsells, M., Recasens, M., Esteve, E., Casamitjana, R., Ricart, W., Fernandez-Real, J.M., 2006. Serum visfatin increases with progressive beta-cell deterioration. Diabetes 55, 2871−2875.

Lowell, B.B., S-Susulic, V., Hamann, A., Lawitts, J.A., Himms-Hagen, J., Boyer, B.B., Kozak, L.P., Flier, J.S., 1993. Development of obesity in transgenic mice after genetic ablation of brown adipose tissue. Nature 366, 740−742.

Luo, L., Liu, M., 2016. Adipose tissue in control of metabolism. J. Endocrinol. 231, R77−R99.

Luo, Y., Ye, S., Chen, X., Gong, F., Lu, W., Li, X., 2017. Rush to the fire: FGF21 extinguishes metabolic stress, metaflammation and tissue damage. Cytokine. Growth. Factor Rev.

Lynes, M.D., Tseng, Y.H., 2017. Deciphering adipose tissue heterogeneity. Ann. N. Y. Acad. Sci.

Mantzoros, C.S., Magkos, F., Brinkoetter, M., Sienkiewicz, E., Dardeno, T.A., Kim, S.Y., Hamnvik, O.P., Koniaris, A., 2011. Leptin in human physiology and pathophysiology. Am. J. Physiol. Endocrinol. Metab. 301, E567−E584.

Maratos-Flier, E., 2017. Fatty liver and FGF21 physiology. Exp. Cell Res. 360, 2−5.

Marchington, J.M., Mattacks, C.A., Pond, C.M., 1989. Adipose tissue in the mammalian heart and pericardium: structure, foetal development and biochemical properties. Comp. Biochem. Physiol. B 94, 225−232.

Marguet, D., Baggio, L., Kobayashi, T., Bernard, A.M., Pierres, M., Nielsen, P.F., Ribel, U., Watanabe, T., Drucker, D.J., Wagtmann, N., 2000. Enhanced insulin secretion and improved glucose tolerance in mice lacking CD26. Proc. Natl. Acad. Sci. U. S. A. 97, 6874—6879.

Matloch, Z., Kotulak, T., Haluzik, M., 2016. The role of epicardial adipose tissue in heart disease. Physiol. Res. 65, 23—32.

Mignatti, P., Morimoto, T., Rifkin, D.B., 1992. Basic fibroblast growth factor, a protein devoid of secretory signal sequence, is released by cells via a pathway independent of the endoplasmic reticulum-Golgi complex. J. Cell Physiol. 151, 81—93.

Miller, R.A., Chu, Q., Le Lay, J., Scherer, P.E., Ahima, R.S., Kaestner, K.H., Foretz, M., Viollet, B., Birnbaum, M.J., 2011. Adiponectin suppresses gluconeogenic gene expression in mouse hepatocytes independent of LKB1-AMPK signaling. J. Clin. Invest. 121, 2518—2528.

Muenzner, M., Tuvia, N., Deutschmann, C., Witte, N., Tolkachov, A., Valai, A., Henze, A., Sander, L.E., Raila, J., Schupp, M., 2013. Retinol-binding protein 4 and its membrane receptor STRA6 control adipogenesis by regulating cellular retinoid homeostasis and retinoic acid receptor alpha activity. Mol. Cell Biol. 33, 4068—4082.

Murawska-Cialowicz, E., 2017. Adipose tissue - morphological and biochemical characteristic of different depots. Postepy Hig. Med. Dosw. 71, 466—484.

Nava, E., Llorens, S., 2016. The paracrine control of vascular motion. A historical perspective. Pharmacol. Res. 113, 125—145.

Nickel, W., 2003. The mystery of nonclassical protein secretion. A current view on cargo proteins and potential export routes. Eur. J. Biochem. 270, 2109—2119.

Norseen, J., Hosooka, T., Hammarstedt, A., Yore, M.M., Kant, S., Aryal, P., Kiernan, U.A., Phillips, D.A., Maruyama, H., Kraus, B.J., Usheva, A., Davis, R.J., Smith, U., Kahn, B.B., 2012. Retinol-binding protein 4 inhibits insulin signaling in adipocytes by inducing proinflammatory cytokines in macrophages through a c-Jun N-terminal kinase- and toll-like receptor 4-dependent and retinol-independent mechanism. Mol. Cell Biol. 32, 2010—2019.

Nosalski, R., Guzik, T.J., 2017. Perivascular adipose tissue inflammation in vascular disease. Br. J. Pharmacol. 174, 3496—3513.

Ohno, H., Shinoda, K., Spiegelman, B.M., Kajimura, S., 2012. PPARgamma agonists induce a white-to-brown fat conversion through stabilization of PRDM16 protein. Cell Metab. 15, 395—404.

Oita, R.C., Ferdinando, D., Wilson, S., Bunce, C., Mazzatti, D.J., 2010. Visfatin induces oxidative stress in differentiated C2C12 myotubes in an Akt- and MAPK-independent, NFkB-dependent manner. Pflugers Arch. 459, 619—630.

Ouchi, N., Higuchi, A., Ohashi, K., Oshima, Y., Gokce, N., Shibata, R., Akasaki, Y., Shimono, A., Walsh, K., 2010. Sfrp5 is an anti-inflammatory adipokine that modulates metabolic dysfunction in obesity. Science 329, 454—457.

Ouedraogo, R., Gong, Y., Berzins, B., Wu, X., Mahadev, K., Hough, K., Chan, L., Goldstein, B.J., Scalia, R., 2007. Adiponectin deficiency increases leukocyte-endothelium interactions via upregulation of endothelial cell adhesion molecules in vivo. J. Clin. Invest. 117, 1718—1726.

Ouwens, D.M., Sell, H., Greulich, S., Eckel, J., 2010. The role of epicardial and perivascular adipose tissue in the pathophysiology of cardiovascular disease. J. Cell Mol. Med. 14, 2223—2234.

Pal, M., Febbraio, M.A., Whitham, M., 2014. From cytokine to myokine: the emerging role of interleukin-6 in metabolic regulation. Immunol. Cell Biol. 92, 331—339.

Palade, G., 1975. Intracellular aspects of the process of protein synthesis. Science 189, 867.

Park, J.R., Jung, J.W., Lee, Y.S., Kang, K.S., 2008. The roles of Wnt antagonists Dkk1 and sFRP4 during adipogenesis of human adipose tissue-derived mesenchymal stem cells. Cell Prolif. 41, 859−874.

Patel, V.B., Shah, S., Verma, S., Oudit, G.Y., 2017. Epicardial adipose tissue as a metabolic transducer: role in heart failure and coronary artery disease. Heart Fail Rev.

Peschechera, A., Eckel, J., 2013. "Browning" of adipose tissue−regulation and therapeutic perspectives. Arch. Physiol. Biochem. 119, 151−160.

Petrovic, N., Walden, T.B., Shabalina, I.G., Timmons, J.A., Cannon, B., Nedergaard, J., 2010. Chronic peroxisome proliferator-activated receptor gamma (PPARgamma) activation of epididymally derived white adipocyte cultures reveals a population of thermogenically competent, UCP1-containing adipocytes molecularly distinct from classic brown adipocytes. J. Biol. Chem. 285, 7153−7164.

Polyzos, S.A., Kountouras, J., Mantzoros, C.S., 2016. Adipokines in nonalcoholic fatty liver disease. Metabolism 65, 1062−1079.

Rakatzi, I., Mueller, H., Ritzeler, O., Tennagels, N., Eckel, J., 2004. Adiponectin counteracts cytokine- and fatty acid-induced apoptosis in the pancreatic beta-cell line INS-1. Diabetologia 47, 249−258.

Ramirez, J.G., O'Malley, E.J., Ho, W.S.V., 2017. Pro-contractile effects of perivascular fat in health and disease. Br. J. Pharmacol. 174, 3482−3495.

Raschke, S., Eckel, J., 2013. Adipo-myokines: two sides of the same coin−mediators of inflammation and mediators of exercise. Mediat. Inflamm. 2013, 320724.

Revollo, J.R., Korner, A., Mills, K.F., Satoh, A., Wang, T., Garten, A., Dasgupta, B., Sasaki, Y., Wolberger, C., Townsend, R.R., Milbrandt, J., Kiess, W., Imai, S., 2007. Nampt/PBEF/Visfatin regulates insulin secretion in beta cells as a systemic NAD biosynthetic enzyme. Cell Metab. 6, 363−375.

Rhee, S.D., Sung, Y.Y., Jung, W.H., Cheon, H.G., 2008. Leptin inhibits rosiglitazone-induced adipogenesis in murine primary adipocytes. Mol. Cell Endocrinol. 294, 61−69.

Richard, D., 2007. Energy expenditure: a critical determinant of energy balance with key hypothalamic controls. Minerva Endocrinol. 32, 173−183.

Rietdorf, K., MacQueen, H., 2017. Investigating interactions between epicardial adipose tissue and cardiac myocytes: what can we learn from different approaches? Br. J. Pharmacol. 174, 3542−3560.

Rittig, K., Dolderer, J.H., Balletshofer, B., Machann, J., Schick, F., Meile, T., Kuper, M., Stock, U.A., Staiger, H., Machicao, F., Schaller, H.E., Konigsrainer, A., Haring, H.U., Siegel-Axel, D.I., 2012. The secretion pattern of perivascular fat cells is different from that of subcutaneous and visceral fat cells. Diabetologia 55, 1514−1525.

Robertson, S.A., Leinninger, G.M., Myers Jr., M.G., 2008. Molecular and neural mediators of leptin action. Physiol. Behav. 94, 637−642.

Roden, M., Stingl, H., Chandramouli, V., Schumann, W.C., Hofer, A., Landau, B.R., Nowotny, P., Waldhausl, W., Shulman, G.I., 2000. Effects of free fatty acid elevation on postabsorptive endogenous glucose production and gluconeogenesis in humans. Diabetes 49, 701−707.

Rohrborn, D., Eckel, J., Sell, H., 2014. Shedding of dipeptidyl peptidase 4 is mediated by metalloproteases and up-regulated by hypoxia in human adipocytes and smooth muscle cells. FEBS Lett. 588, 3870−3877.

Rohrborn, D., Wronkowitz, N., Eckel, J., 2015. DPP4 in diabetes. Front. Immunol. 6, 386.

Romacho, T., Elsen, M., Rohrborn, D., Eckel, J., 2014. Adipose tissue and its role in organ crosstalk. Acta Physiol. (Oxf) 210, 733−753.

Romacho, T., Sanchez-Ferrer, C.F., Peiro, C., 2013. Visfatin/Nampt: an adipokine with cardiovascular impact. Mediat. Inflamm. 2013, 946427.

Rosell, M., Kaforou, M., Frontini, A., Okolo, A., Chan, Y.W., Nikolopoulou, E., Millership, S., Fenech, M.E., MacIntyre, D., Turner, J.O., Moore, J.D., Blackburn, E., Gullick, W.J., Cinti, S., Montana, G., Parker, M.G., Christian, M., 2014. Brown and white adipose tissues: intrinsic differences in gene expression and response to cold exposure in mice. Am. J. Physiol. Endocrinol. Metab. 306, E945−E964.

Rothwell, N.J., Stock, M.J., 1982. Effect of chronic food restriction on energy balance, thermogenic capacity, and brown-adipose-tissue activity in the rat. Biosci. Rep. 2, 543−549.

Sacks, H.S., Fain, J.N., Holman, B., Cheema, P., Chary, A., Parks, F., Karas, J., Optican, R., Bahouth, S.W., Garrett, E., Wolf, R.Y., Carter, R.A., Robbins, T., Wolford, D., Samaha, J., 2009. Uncoupling protein-1 and related messenger ribonucleic acids in human epicardial and other adipose tissues: epicardial fat functioning as brown fat. J. Clin. Endocrinol. Metab. 94, 3611−3615.

Saito, M., Okamatsu-Ogura, Y., Matsushita, M., Watanabe, K., Yoneshiro, T., Nio-Kobayashi, J., Iwanaga, T., Miyagawa, M., Kameya, T., Nakada, K., Kawai, Y., Tsujisaki, M., 2009. High incidence of metabolically active brown adipose tissue in healthy adult humans: effects of cold exposure and adiposity. Diabetes 58, 1526−1531.

Salazar, J., Luzardo, E., Mejias, J.C., Rojas, J., Ferreira, A., Rivas-Rios, J.R., Bermudez, V., 2016. Epicardial fat: physiological, pathological, and therapeutic implications. Cardiol. Res. Pract. 2016, 1291537.

Salminen, A., Kaarniranta, K., Kauppinen, A., 2017a. Integrated stress response stimulates FGF21 expression: systemic enhancer of longevity. Cell Signal 40, 10−21.

Salminen, A., Kaarniranta, K., Kauppinen, A., 2017b. Regulation of longevity by FGF21: interaction between energy metabolism and stress responses. Ageing Res. Rev. 37, 79−93.

Salminen, A., Kauppinen, A., Kaarniranta, K., 2017c. FGF21 activates AMPK signaling: impact on metabolic regulation and the aging process. J. Mol. Med. Berl. 95, 123−131.

Sarjeant, K., Stephens, J.M., 2012. Adipogenesis. Cold. Spring. Harb. Perspect. Biol. 4, a008417.

Sartipy, P., Loskutoff, D.J., 2003. Monocyte chemoattractant protein 1 in obesity and insulin resistance. Proc. Natl. Acad. Sci. U. S. A. 100, 7265−7270.

Schafer, K., Drosos, I., Konstantinides, S., 2017. Perivascular adipose tissue: epiphenomenon or local risk factor? Int. J. Obes. 41, 1311−1323.

Schafer, T., Zentgraf, H., Zehe, C., Brugger, B., Bernhagen, J., Nickel, W., 2004. Unconventional secretion of fibroblast growth factor 2 is mediated by direct translocation across the plasma membrane of mammalian cells. J. Biol. Chem. 279, 6244−6251.

Scherer, P.E., Williams, S., Fogliano, M., Baldini, G., Lodish, H.F., 1995. A novel serum protein similar to C1q, produced exclusively in adipocytes. J. Biol. Chem. 270, 26746−26749.

Schulz, T.J., Huang, T.L., Tran, T.T., Zhang, H., Townsend, K.L., Shadrach, J.L., Cerletti, M., McDougall, L.E., Giorgadze, N., Tchkonia, T., Schrier, D., Falb, D., Kirkland, J.L., Wagers, A.J., Tseng, Y.H., 2011. Identification of inducible brown adipocyte progenitors residing in skeletal muscle and white fat. Proc. Natl. Acad. Sci. U. S. A. 108, 143−148.

Schulz, T.J., Tseng, Y.H., 2009. Emerging role of bone morphogenetic proteins in adipogenesis and energy metabolism. Cytokine Growth Factor Rev. 20, 523−531.

Seale, P., Bjork, B., Yang, W., Kajimura, S., Chin, S., Kuang, S., Scime, A., Devarakonda, S., Conroe, H.M., Erdjument-Bromage, H., Tempst, P., Rudnicki, M.A., Beier, D.R., Spiegelman, B.M., 2008. PRDM16 controls a brown fat/skeletal muscle switch. Nature 454, 961−967.

Sell, H., Bluher, M., Kloting, N., Schlich, R., Willems, M., Ruppe, F., Knoefel, W.T., Dietrich, A., Fielding, B.A., Arner, P., Frayn, K.N., Eckel, J., 2013. Adipose dipeptidyl peptidase-4 and obesity: correlation with insulin resistance and depot-specific release from adipose tissue in vivo and in vitro. Diabetes Care 36, 4083−4090.

Sell, H., Dietze-Schroeder, D., Eckel, J., 2006a. The adipocyte-myocyte axis in insulin resistance. Trends Endocrinol. Metab. 17, 416−422.

Sell, H., Dietze-Schroeder, D., Kaiser, U., Eckel, J., 2006b. Monocyte chemotactic protein-1 is a potential player in the negative cross-talk between adipose tissue and skeletal muscle. Endocrinology 147, 2458−2467.

Sell, H., Eckel, J., 2007. Regulation of retinol binding protein 4 production in primary human adipocytes by adiponectin, troglitazone and TNF-alpha. Diabetologia 50, 2221−2223.

Sell, H., Habich, C., Eckel, J., 2012. Adaptive immunity in obesity and insulin resistance. Nat. Rev. Endocrinol. 8, 709−716.

Sell, H., Laurencikiene, J., Taube, A., Eckardt, K., Cramer, A., Horrighs, A., Arner, P., Eckel, J., 2009. Chemerin is a novel adipocyte-derived factor inducing insulin resistance in primary human skeletal muscle cells. Diabetes 58, 2731−2740.

Seufert, J., 2004. Leptin effects on pancreatic beta-cell gene expression and function. Diabetes 53 (Suppl 1), S152−S158.

Shah, Z., Pineda, C., Kampfrath, T., Maiseyeu, A., Ying, Z., Racoma, I., Deiuliis, J., Xu, X., Sun, Q., Moffatt-Bruce, S., Villamena, F., Rajagopalan, S., 2011. Acute DPP-4 inhibition modulates vascular tone through GLP-1 independent pathways. Vasc. Pharmacol. 55, 2−9.

Siegel-Axel, D.I., Haring, H.U., 2016. Perivascular adipose tissue: an unique fat compartment relevant for the cardiometabolic syndrome. Rev. Endocr. Metab. Disord. 17, 51−60.

Sindhu, S., Thomas, R., Shihab, P., Sriraman, D., Behbehani, K., Ahmad, R., 2015. Obesity is a positive modulator of IL-6R and IL-6 expression in the subcutaneous adipose tissue: significance for metabolic inflammation. PLoS One 10, e0133494.

Singh, P., Hoffmann, M., Wolk, R., Shamsuzzaman, A.S., Somers, V.K., 2007. Leptin induces C-reactive protein expression in vascular endothelial cells. Arterioscler. Thromb. Vasc. Biol. 27, e302−e307.

Singh, P., Peterson, T.E., Barber, K.R., Kuniyoshi, F.S., Jensen, A., Hoffmann, M., Shamsuzzaman, A.S., Somers, V.K., 2010. Leptin upregulates the expression of plasminogen activator inhibitor-1 in human vascular endothelial cells. Biochem. Biophys. Res. Commun. 392, 47−52.

Singh, P., Peterson, T.E., Sert-Kuniyoshi, F.H., Jensen, M.D., Somers, V.K., 2011. Leptin upregulates caveolin-1 expression: implications for development of atherosclerosis. Atherosclerosis 217, 499−502.

Skurk, T., Alberti-Huber, C., Herder, C., Hauner, H., 2007. Relationship between adipocyte size and adipokine expression and secretion. J. Clin. Endocrinol. Metab. 92, 1023−1033.

Soedling, H., Hodson, D.J., Adrianssens, A.E., Gribble, F.M., Reimann, F., Trapp, S., Rutter, G.A., 2015. Limited impact on glucose homeostasis of leptin receptor deletion from insulin- or proglucagon-expressing cells. Mol. Metab. 4, 619−630.

Sonoda, J., Chen, M.Z., Baruch, A., 2017. FGF21-receptor agonists: an emerging therapeutic class for obesity-related diseases. Horm. Mol. Biol. Clin. Investig. 30.

Spalding, K.L., Arner, E., Westermark, P.O., Bernard, S., Buchholz, B.A., Bergmann, O., Blomqvist, L., Hoffstedt, J., Naslund, E., Britton, T., Concha, H., Hassan, M., Ryden, M., Frisen, J., Arner, P., 2008. Dynamics of fat cell turnover in humans. Nature 453, 783–787.

Spinnler, R., Gorski, T., Stolz, K., Schuster, S., Garten, A., Beck-Sickinger, A.G., Engelse, M.A., de Koning, E.J., Korner, A., Kiess, W., Maedler, K., 2013. The adipocytokine Nampt and its product NMN have no effect on beta-cell survival but potentiate glucose stimulated insulin secretion. PLoS One 8, e54106.

Spiroglou, S.G., Kostopoulos, C.G., Varakis, J.N., Papadaki, H.H., 2010. Adipokines in periaortic and epicardial adipose tissue: differential expression and relation to atherosclerosis. J. Atheroscler. Thromb. 17, 115–130.

Staiger, K., Stefan, N., Staiger, H., Brendel, M.D., Brandhorst, D., Bretzel, R.G., Machicao, F., Kellerer, M., Stumvoll, M., Fritsche, A., Haring, H.U., 2005. Adiponectin is functionally active in human islets but does not affect insulin secretory function or beta-cell lipoapoptosis. J. Clin. Endocrinol. Metab. 90, 6707–6713.

Stanford, K.I., Goodyear, L.J., 2016. Exercise regulation of adipose tissue. Adipocyte 5, 153–162.

Stanford, K.I., Middelbeek, R.J., Townsend, K.L., An, D., Nygaard, E.B., Hitchcox, K.M., Markan, K.R., Nakano, K., Hirshman, M.F., Tseng, Y.H., Goodyear, L.J., 2013. Brown adipose tissue regulates glucose homeostasis and insulin sensitivity. J. Clin. Invest. 123, 215–223.

Stern, J.H., Rutkowski, J.M., Scherer, P.E., 2016. Adiponectin, leptin, and fatty acids in the maintenance of metabolic homeostasis through adipose tissue crosstalk. Cell Metab. 23, 770–784.

Strowski, M.Z., 2017. Impact of FGF21 on glycemic control. Horm. Mol. Biol. Clin. Investig. 30.

Sweeney, G., 2010. Cardiovascular effects of leptin. Nat. Rev. Cardiol. 7, 22–29.

Swifka, J., Weiss, J., Addicks, K., Eckel, J., Rosen, P., 2008. Epicardial fat from Guinea pig: a model to study the paracrine network of interactions between epicardial fat and myocardium? Cardiovasc. Drugs Ther. 22, 107–114.

Tang, Q.Q., Lane, M.D., 2012. Adipogenesis: from stem cell to adipocyte. Annu. Rev. Biochem. 81, 715–736.

Taniguchi, C.M., Emanuelli, B., Kahn, C.R., 2006. Critical nodes in signalling pathways: insights into insulin action. Nat. Rev. Mol. Cell Biol. 7, 85–96.

Taube, A., Schlich, R., Sell, H., Eckardt, K., Eckel, J., 2012. Inflammation and metabolic dysfunction: links to cardiovascular diseases. Am. J. Physiol. Heart Circ. Physiol. 302, H2148–H2165.

Timmons, J.A., Wennmalm, K., Larsson, O., Walden, T.B., Lassmann, T., Petrovic, N., Hamilton, D.L., Gimeno, R.E., Wahlestedt, C., Baar, K., Nedergaard, J., Cannon, B., 2007. Myogenic gene expression signature establishes that brown and white adipocytes originate from distinct cell lineages. Proc. Natl. Acad. Sci. U. S. A. 104, 4401–4406.

Townsend, K.L., Tseng, Y.H., 2015. Of mice and men: novel insights regarding constitutive and recruitable brown adipocytes. Int. J. Obes. Suppl. 5, S15–S20.

Trayhurn, P., Jones, P.M., McGuckin, M.M., Goodbody, A.E., 1982. Effects of overfeeding on energy balance and brown fat thermogenesis in obese (ob/ob) mice. Nature 295, 323–325.

Tsutsumi, C., Okuno, M., Tannous, L., Piantedosi, R., Allan, M., Goodman, D.S., Blaner, W.S., 1992. Retinoids and retinoid-binding protein expression in rat adipocytes. J. Biol. Chem. 267, 1805–1810.

Turer, A.T., Scherer, P.E., 2012. Adiponectin: mechanistic insights and clinical implications. Diabetologia 55, 2319−2326.

Uysal, K.T., Wiesbrock, S.M., Marino, M.W., Hotamisligil, G.S., 1997. Protection from obesity-induced insulin resistance in mice lacking TNF-alpha function. Nature 389, 610−614.

van Marken Lichtenbelt, W.D., Vanhommerig, J.W., Smulders, N.M., Drossaerts, J.M., Kemerink, G.J., Bouvy, N.D., Schrauwen, P., Teule, G.J., 2009. Cold-activated brown adipose tissue in healthy men. N. Engl. J. Med. 360, 1500−1508.

Vanella, L., Sodhi, K., Kim, D.H., Puri, N., Maheshwari, M., Hinds, T.D., Bellner, L., Goldstein, D., Peterson, S.J., Shapiro, J.I., Abraham, N.G., 2013. Increased heme-oxygenase 1 expression in mesenchymal stem cell-derived adipocytes decreases differentiation and lipid accumulation via upregulation of the canonical Wnt signaling cascade. Stem Cell Res. Ther. 4, 28.

Venteclef, N., Guglielmi, V., Balse, E., Gaborit, B., Cotillard, A., Atassi, F., Amour, J., Leprince, P., Dutour, A., Clement, K., Hatem, S.N., 2015. Human epicardial adipose tissue induces fibrosis of the atrial myocardium through the secretion of adipo-fibrokines. Eur. Heart J. 36, 795−805.

Verhagen, S.N., Visseren, F.L., 2011. Perivascular adipose tissue as a cause of atherosclerosis. Atherosclerosis 214, 3−10.

Verma, S., Buchanan, M.R., Anderson, T.J., 2003. Endothelial function testing as a biomarker of vascular disease. Circulation 108, 2054−2059.

Villarroya, J., Cereijo, R., Villarroya, F., 2013. An endocrine role for brown adipose tissue? Am. J. Physiol. Endocrinol. Metab. 305, E567−E572.

Virtanen, K.A., Lidell, M.E., Orava, J., Heglind, M., Westergren, R., Niemi, T., Taittonen, M., Laine, J., Savisto, N.J., Enerback, S., Nuutila, P., 2009. Functional brown adipose tissue in healthy adults. N. Engl. J. Med. 360, 1518−1525.

Wajchenberg, B.L., 2007. beta-cell failure in diabetes and preservation by clinical treatment. Endocr. Rev. 28, 187−218.

Waki, H., Park, K.W., Mitro, N., Pei, L., Damoiseaux, R., Wilpitz, D.C., Reue, K., Saez, E., Tontonoz, P., 2007. The small molecule harmine is an antidiabetic cell-type-specific regulator of PPARgamma expression. Cell Metab. 5, 357−370.

Wang, G.X., Zhao, X.Y., Lin, J.D., 2015. The brown fat secretome: metabolic functions beyond thermogenesis. Trends Endocrinol. Metab. 26, 231−237.

Wang, G.X., Zhao, X.Y., Meng, Z.X., Kern, M., Dietrich, A., Chen, Z., Cozacov, Z., Zhou, D., Okunade, A.L., Su, X., Li, S., Bluher, M., Lin, J.D., 2014. The brown fat-enriched secreted factor Nrg4 preserves metabolic homeostasis through attenuation of hepatic lipogenesis. Nat. Med. 20, 1436−1443.

Wang, X., Guo, Z., Zhu, Z., Bao, Y., Yang, B., 2016. Epicardial fat tissue in patients with psoriasis:a systematic review and meta-analysis. Lipids Health Dis. 15, 103.

Wang, Y., Lam, K.S., Yau, M.H., Xu, A., 2008. Post-translational modifications of adiponectin: mechanisms and functional implications. Biochem. J. 409, 623−633.

Weigert, C., Hennige, A.M., Brodbeck, K., Haring, H.U., Schleicher, E.D., 2005. Interleukin-6 acts as insulin sensitizer on glycogen synthesis in human skeletal muscle cells by phosphorylation of Ser473 of Akt. Am. J. Physiol. Endocrinol. Metab. 289, E251−E257.

White, M.F., 2003. Insulin signaling in health and disease. Science 302, 1710−1711.

Wijers, S.L., Schrauwen, P., Saris, W.H., van Marken Lichtenbelt, W.D., 2008. Human skeletal muscle mitochondrial uncoupling is associated with cold induced adaptive thermogenesis. PLoS One 3, e1777.

Wijesekara, N., Krishnamurthy, M., Bhattacharjee, A., Suhail, A., Sweeney, G., Wheeler, M.B., 2010. Adiponectin-induced ERK and Akt phosphorylation protects against pancreatic beta cell apoptosis and increases insulin gene expression and secretion. J. Biol. Chem. 285, 33623–33631.

William Jr., W.N., Ceddia, R.B., Curi, R., 2002. Leptin controls the fate of fatty acids in isolated rat white adipocytes. J. Endocrinol. 175, 735–744.

Williams, I.L., Wheatcroft, S.B., Shah, A.M., Kearney, M.T., 2002. Obesity, atherosclerosis and the vascular endothelium: mechanisms of reduced nitric oxide bioavailability in obese humans. Int. J. Obes. Relat. Metab. Disord. 26, 754–764.

Wronkowitz, N., Romacho, T., Sell, H., Eckel, J., 2014. Adipose tissue dysfunction and inflammation in cardiovascular disease. Front. Horm. Res. 43, 79–92.

Wu, J., Bostrom, P., Sparks, L.M., Ye, L., Choi, J.H., Giang, A.H., Khandekar, M., Virtanen, K.A., Nuutila, P., Schaart, G., Huang, K., Tu, H., van Marken Lichtenbelt, W.D., Hoeks, J., Enerback, S., Schrauwen, P., Spiegelman, B.M., 2012. Beige adipocytes are a distinct type of thermogenic fat cell in mouse and human. Cell 150, 366–376.

Wu, Y., Zhang, A., Hamilton, D.J., Deng, T., 2017. Epicardial fat in the maintenance of cardiovascular health. Methodist. Debakey Cardiovasc. J. 13, 20–24.

Xia, N., Li, H., 2017. The role of perivascular adipose tissue in obesity-induced vascular dysfunction. Br. J. Pharmacol. 174, 3425–3442.

Xiao, L., Sobue, T., Esliger, A., Kronenberg, M.S., Coffin, J.D., Doetschman, T., Hurley, M.M., 2010. Disruption of the Fgf2 gene activates the adipogenic and suppresses the osteogenic program in mesenchymal marrow stromal stem cells. Bone 47, 360–370.

Xie, L., Boyle, D., Sanford, D., Scherer, P.E., Pessin, J.E., Mora, S., 2006. Intracellular trafficking and secretion of adiponectin is dependent on GGA-coated vesicles. J. Biol. Chem. 281, 7253–7259.

Xie, L., O'Reilly, C.P., Chapes, S.K., Mora, S., 2008. Adiponectin and leptin are secreted through distinct trafficking pathways in adipocytes. Biochim. Biophys. Acta 1782, 99–108.

Xu, A., Wang, Y., Keshaw, H., Xu, L.Y., Lam, K.S., Cooper, G.J., 2003. The fat-derived hormone adiponectin alleviates alcoholic and nonalcoholic fatty liver diseases in mice. J. Clin. Invest. 112, 91–100.

Yamauchi, T., Kamon, J., Ito, Y., Tsuchida, A., Yokomizo, T., Kita, S., Sugiyama, T., Miyagishi, M., Hara, K., Tsunoda, M., Murakami, K., Ohteki, T., Uchida, S., Takekawa, S., Waki, H., Tsuno, N.H., Shibata, Y., Terauchi, Y., Froguel, P., Tobe, K., Koyasu, S., Taira, K., Kitamura, T., Shimizu, T., Nagai, R., Kadowaki, T., 2003. Cloning of adiponectin receptors that mediate antidiabetic metabolic effects. Nature 423, 762–769.

Yamauchi, T., Kamon, J., Minokoshi, Y., Ito, Y., Waki, H., Uchida, S., Yamashita, S., Noda, M., Kita, S., Ueki, K., Eto, K., Akanuma, Y., Froguel, P., Foufelle, F., Ferre, P., Carling, D., Kimura, S., Nagai, R., Kahn, B.B., Kadowaki, T., 2002. Adiponectin stimulates glucose utilization and fatty-acid oxidation by activating AMP-activated protein kinase. Nat. Med. 8, 1288–1295.

Yang, Q., Graham, T.E., Mody, N., Preitner, F., Peroni, O.D., Zabolotny, J.M., Kotani, K., Quadro, L., Kahn, B.B., 2005. Serum retinol binding protein 4 contributes to insulin resistance in obesity and type 2 diabetes. Nature 436, 356–362.

Yao-Borengasser, A., Varma, V., Bodles, A.M., Rasouli, N., Phanavanh, B., Lee, M.J., Starks, T., Kern, L.M., Spencer 3rd, H.J., Rashidi, A.A., McGehee Jr., R.E., Fried, S.K., Kern, P.A., 2007. Retinol binding protein 4 expression in humans: relationship to insulin resistance, inflammation, and response to pioglitazone. J. Clin. Endocrinol. Metab. 92, 2590−2597.

Youn, B.S., Kloting, N., Kratzsch, J., Lee, N., Park, J.W., Song, E.S., Ruschke, K., Oberbach, A., Fasshauer, M., Stumvoll, M., Bluher, M., 2008. Serum vaspin concentrations in human obesity and type 2 diabetes. Diabetes 57, 372−377.

Zaborska, K.E., Wareing, M., Austin, C., 2017. Comparisons between perivascular adipose tissue and the endothelium in their modulation of vascular tone. Br. J. Pharmacol. 174, 3388−3397.

Zacharogianni, M., Rabouille, C., 2013. Trafficking along the secretory pathway in Drosophila cell line and tissues: a light and electron microscopy approach. Methods Cell Biol. 118, 35−49.

Zamani, N., Brown, C.W., 2011. Emerging roles for the transforming growth factor-{beta} superfamily in regulating adiposity and energy expenditure. Endocr. Rev. 32, 387−403.

Zeng, W., Pirzgalska, R.M., Pereira, M.M., Kubasova, N., Barateiro, A., Seixas, E., Lu, Y.H., Kozlova, A., Voss, H., Martins, G.G., Friedman, J.M., Domingos, A.I., 2015. Sympathetic neuro-adipose connections mediate leptin-driven lipolysis. Cell 163, 84−94.

Zhang, Y., Proenca, R., Maffei, M., Barone, M., Leopold, L., Friedman, J.M., 1994. Positional cloning of the mouse obese gene and its human homologue. Nature 372, 425−432.

Zingaretti, M.C., Crosta, F., Vitali, A., Guerrieri, M., Frontini, A., Cannon, B., Nedergaard, J., Cinti, S., 2009. The presence of UCP1 demonstrates that metabolically active adipose tissue in the neck of adult humans truly represents brown adipose tissue. FASEB J. 23, 3113−3120.

Ziouzenkova, O., Plutzky, J., 2008. Retinoid metabolism and nuclear receptor responses: new insights into coordinated regulation of the PPAR-RXR complex. FEBS Lett. 582, 32−38.

Skeletal Muscle: A Novel Secretory Organ

3

CHAPTER OUTLINE

As outlined in detail in Chapter 2, adipose tissue is now widely accepted to represent a major endocrine organ with an exceptional capacity to release a host of bioactive molecules. In contrast, the role and contribution of skeletal muscle to the complex scenario of organ crosstalk has remained largely unexplored, although it has been speculated for a longtime that a so-called "exercise factor" would be able to communicate between contracting muscle and other organs. Driven by the research on adipose tissue and the new results on adipokines and their important impact for metabolic homeostasis, more recent work started to focus on skeletal muscle as a potential endocrine organ, and the myokine concept was developed and suggested as a conceptual framework underlying the crosstalk from skeletal muscle to other organs (Pedersen, 2013). Skeletal muscle as part of the organ communication network is of special interest for several reasons: First, in nonobese adult humans, skeletal muscle represents the most abundant tissue, and therefore, the muscular secretome may substantially contribute to the level of circulating factors being involved in organ crosstalk. Second, skeletal muscle has a high degree of plasticity and adapts to a variety of stimuli that play a key role in the regulation of systemic metabolism. Finally, and perhaps most importantly, the muscular secretome is tightly regulated by muscle contraction. The latter is linked to physical exercise, which plays a key role for well-being and prevents onset and development of numerous diseases.

The Cellular Secretome and Organ Crosstalk. https://doi.org/10.1016/B978-0-12-809518-8.00003-9

Although the health-promoting effects of physical exercise are well-known for a long time, the molecular basis of these effects has remained poorly understood. Physical exercise is now considered as a major preventive public health intervention, as it is well-known to reduce all-cause mortality and major chronic diseases including type 2 diabetes, cardiovascular disease, and different forms of cancer, dementia, and depression (Eckardt et al., 2014). Physical activity is often considered as a life-style factor that affects body weight, energy intake, and fat mass. However, detailed studies on the effects of physical inactivity have shown the presence of chronic inflammation in the absence of obesity, pointing to an independent role of muscle contraction in the control of systemic metabolism. The myokine concept has opened a new window to understand the molecular consequences of physical exercise, to improve existing strategies for exercise interventions, and to identify novel target molecules that mediate the beneficial effects of regular physical activity. In this chapter, the skeletal muscle secretome and the myokine concept is presented. Old and novel myokines are discussed and evaluated, with emphasis on the critical and controversial findings in the literature. Finally, we will focus on the muscle-to-fat crosstalk as a paradigm of bidirectional communication between two organs and the muscle—bone crosstalk, a new and intriguing aspect of organ communication.

THE MYOKINE CONCEPT AND THE SKELETAL MUSCLE SECRETOME

When considering the effects of exercise, it is important to differentiate between acute exercise protocols and exercise training, which consists of repeated bouts of exercise that lead to an adaptive response, in contrast to the stresslike response produced by acute exercise. As pointed out before, exercise training has deep impact on systemic metabolism, and muscle has even been termed a gene regulatory organ (Karstoft and Pedersen, 2016). Adaptation to chronic exercise leads to increased muscle oxidative capacity and improved insulin sensitivity. However, these adaptations extend to other distinct organs such as adipose tissue, the vasculature, and even to bone (Tagliaferri et al., 2015). It is now widely accepted that these distal regulatory effects are mediated by peptides or proteins released from skeletal muscle cells. Many of these molecules, collectively named myokines, belong to the cytokine family of interleukins (ILs), with IL-6 representing the first identified protein being released to the circulation after exercise (Steensberg et al., 2002). Thus, IL-6 can be considered as the first myokine and still represents the gold standard of an exercise factor (see below). 15 years later, owing to highly sensitive technologies and sophisticated approaches, hundreds of myokines have been described in the literature with mostly undefined functions. To avoid confusion and to remain on track with the search for physiologically relevant targets, a clear definition of a myokine is mandatory. This is presented in Box 3.1 and the following passage.

> ### BOX 3.1 KEY FACTS ABOUT MYOKINES
> - A myokine is defined as a peptide or protein secreted or released from skeletal muscle cells.
> - Myokines may exert auto-, para-, or endocrine functions.
> - A myokine can be an exercise factor (such as IL-6), but these molecules are released to the circulation and are not necessarily peptides or proteins.
> - Both expression and release of myokines can be regulated by muscle contraction, but this is not a mandatory property of a myokine.

MYOKINES AND EXERCISE FACTORS: A DEFINITION

The humoral nature of exercise factors released by working skeletal muscle was postulated more than 50 years ago (Goldstein, 1961), but the seminal work by Pedersen et al. (2003, 2004), and Pedersen and Febbraio (2012) set the stage for the myokine concept and provided convincing evidence for the existence of an exercise factor, in this case IL-6, which can be considered as the prototype myokine. As outlined in Box 3.1, we define a myokine as a peptide or protein secreted by skeletal muscle cells. This does not imply that these molecules are released to the circulation; instead, many of them (maybe even the majority) remain within skeletal muscle and exert auto- or paracrine functions. Contractile activity of myocytes involves activation of Ca^{++} and other signaling pathways that may regulate transcription and/or secretion of myokines. However, not all myokines are regulated by contractile activity, and we recommend that the release from myocytes is a mandatory criterion for the term "myokine". Unfortunately, the situation is much more complex in that most likely none of the myokines is produced exclusively by myocytes. The best example is the prototype myokine IL-6 itself, which is also a well-known proinflammatory adipokine (see Chapter 2). We propose that the complete set of all peptides and proteins (organokinome) produced by all tissues comprises a limited number of molecules that gain specificity by (1) either a specific signature in the form of covalent modifications in a tissue-specific context, or (2) receptor-mediated processes at the level of target cells. This exciting topic is still unexplored and needs to be addressed in the future. We have used the term "adipomyokines" to emphasize this overlap between adipokines and myokines (Raschke and Eckel, 2013) (for detailed discussion, see Chapter 4).

In contrast to a myokine, an exercise factor is released to the circulation and is not necessarily a protein. Owing to modern metabolomics approaches, a number of metabolites have been identified as potential exercise factors. Compared with the myokine concept, the role of metabolites in mediating muscle-to-organ crosstalk appears much less important. This may be explained by a much higher information capacity of a protein being carried to a specific receptor and signaling pathway at a target cell.

SECRETOME ANALYSIS

The myokine concept outlined above can be considered as the conceptual platform for analyzing and understanding the crosstalk from muscle to a variety of other organs. Secretomic studies to identify myokines released from skeletal muscle or isolated myocytes kept in culture have therefore gained considerable interest, and at present more than 600 myokines have been described (Gorgens et al., 2015). However, the muscle secretome is still insufficiently characterized. Most importantly, in many cases, the available data do not go beyond profiling, and the vast majority of identified myokines is still requiring functional analysis. On top, for many of the hundreds of myokines, we do not know whether they are regulated by exercise and whether they represent exercise factors or function in an autocrine/paracrine fashion. Compared with adipose tissue, the situation is more complex. First, the concentration of myokines released from muscles cells is much lower compared with that from adipocytes. This is certainly compensated in vivo by the higher abundance of skeletal muscle compared with adipose tissue, but complicates the analysis when using cell culture models. Second, appropriate models that mimic muscle contraction, a central feature of muscle function, have only recently been described, specifically for humans.

One recently published approach (Catoire et al., 2014b) is based on the ex vivo analysis of human skeletal muscle biopsies taken before and after acute exercise and exercise training, followed by microarray-based analysis of secreted proteins. This resulted in a short list of 29 putative myokines that were further analyzed by multiplex technologies. Finally, MCP-1 and fractalkine were found to be increased at both the mRNA and the plasma level, and these myokines could play a yet-to-be-defined role in the communication between skeletal muscle and other organs. Several other studies used a screening approach and subsequently validated single myokines. This approach successfully identified angiopoietin-like 4 (Catoire et al., 2014a), IL-7 (Haugen et al., 2010), apelin (Besse-Patin et al., 2014), and several others. An unbiased approach aiming to characterize the secretome of differentiated primary human skeletal muscle cells was recently reported by Lehr et al. (Hartwig et al., 2014). This study combined genomics, three different mass spectrometry methods, and multiplex immunoassays to generate a comprehensive characterization of the entire human muscle secretome. Using this approach, the authors identified 548 proteins in conditioned media of the human skeletal muscle cells (see also Chapter 6). Based on stringent consecutive filtering, 305 proteins were assigned as potential myokines. These data show that skeletal muscle cells have a prominent basic secretory capacity, principally not different from adipocytes. As will be discussed in the next chapter, many of these myokines are also adipokines and are produced by adipose tissue. A key question that needs to be addressed in the future is regarding the regulation of these myokines both at the level of expression and secretion from the myocytes. In this context, an interesting study using mass spectrometry—based proteomics showed that lipid-induced insulin resistance affected nearly one-half of the secreted proteins (Deshmukh et al., 2015). Much more studies will be needed to identify potential modifications of the skeletal muscle secretome owing to physiological or pathophysiological stimuli.

A major physiological stimulus is muscle contraction, and a variety of signal transduction pathways are activated in contracting muscles, depending on exercise intensity and duration. Furthermore, muscle contraction is considered as a key regulator of myokine expression and release. Electrical pulse stimulation (EPS) of cultured skeletal muscle cells turned out to be a major tool for proteomic studies on the contraction-regulated secretome, and application of this model to human cells was instrumental for gaining new insights in the crosstalk between contracting muscle and other organs (Gorgens et al., 2015). Application of EPS as an in vitro exercise model was first reported in 2008 using the murine C2C12 skeletal muscle cell line (Nedachi et al., 2008). These authors showed that EPS for 24 h activated AMP-kinase and induced the expression of several myokines known to be released from contracting muscle in vivo, such as IL-6. A major breakthrough was obtained, when two groups independently (Lambernd et al., 2012; Nikolic et al., 2012) described a model of chronic low-frequency electrical stimulation of human myotubes. These cells show adaptation to the continuous contraction by reorganization of the cytoskeleton, de novo formation of sarcomeric structures, increased glucose uptake and insulin sensitivity, and increased mitochondrial content. Although a variety of EPS protocols are currently used that may explain a number of divergent findings, it is generally accepted that EPS of human muscle cells is a valid model to study contraction-mediated processes and adaptation to exercise training.

Using this approach and an antibody-array technology to analyze the secretory output, a total of 48 myokines was reported to be regulated by muscle contraction including 21 novel contraction-regulated myokines (Raschke et al., 2013a). It is also important to note that myotubes from different donors were found to release a substantially different set of myokines, making it likely that an individual signature of myokines exists. This may have considerable impact for training interventions, and future work will be needed to address this issue. Taken together, the analysis of conditioned media from EPS models has generated additional knowledge on the complex scenario of skeletal muscle as an endocrine organ. However, there are a number of limitations that need to be kept in mind. First, there are several technical issues related to secretome analysis, which was already mentioned in Chapter 2. This includes preparation and processing of conditioned media from muscle cells or explants, the selection of mass spectrometry technologies including quantitative analysis, and the bioinformatic processing of the omics data. These critical issues of applying secretomic analysis to skeletal muscle cells have been summarized recently by Yoon et al. (2012). An additional limitation of the EPS model is certainly the lack of work load in this system and inability to fully mimic the complex scenario of in vivo muscle contraction. Thus, presently hundreds of putative myokines have been identified and described to be regulated at the mRNA or secretory level by differentiation, contraction, or other stimuli. Because many of these myokines have only been described in cell culture models, their relevance for in vivo crosstalk, specifically in humans, remains to be validated. Table 3.1 aims to present a list of

Table 3.1 Selected Myokines With Proven Function in Humans

Myokine	Autocrine/Paracrine Effect	Endocrine Effect
IL-6	Known to increase glucose uptake, glycogen synthesis, fatty acid oxidation, and lipolysis	Increased liver glucose output, augmented glucagon-like peptide-1 secretion, protective effect on the pancreas, and documented crosstalk to skin and bone
IL-8	Positive effects on muscular angiogenesis	Endocrine effects not established
IL-13	Shown to increase glucose uptake and oxidation and glycogen synthesis	Reduced glucose output by the liver
IL-15	Augments fatty acid oxidation	Inhibits lipid accumulation in adipose tissue and stimulates adiponectin secretion
FGF21	Increased glucose uptake	Browning of white fat; increased adiponectin secretion from adipocytes; increased fatty acid oxidation in the liver
ANGPTL4	Prevents local fat overload and directs fatty acids to active skeletal muscle	Endocrine function not yet established

myokines with confirmed function in humans. As can be seen, this list is rather small with only about 5% of the reported number of myokines. It remains to be shown whether the remaining 95% of identified myokines play a role in humans. Many of them most likely do but potentially in a sophisticated autocrine fashion, which is very difficult to analyze. In the following passages, we discuss some molecules that represent typical myokines, although some overlap exists with the secretome of other organs. Some myokines that are also adipokines have also been named "adipomyokines", and these molecules will be presented in the next chapter.

IL-6 AND OTHER INTERLEUKINS

IL-6 is the first and most studied myokine, although there is still a controversial discussion whether it exerts a real myokine function. The major reason is that IL-6 is a very well-known cytokine with documented associations to inflammatory diseases and the development of insulin resistance in adipose tissue and the liver (Catoire and Kersten, 2015). Undoubtedly, the paper by Steensberg et al. (2000) can be considered as a landmark in this research field, as it was the first to report that contracting human skeletal muscle releases IL-6 into the circulation, setting the stage for a new role of skeletal muscle as an endocrine organ. These initial reports were followed by numerous studies on IL-6 to explore its regulation and function as an exercise factor. It has been shown that the exercise-mediated increase in IL-6 mRNA and serum levels is dependent on blood glucose and the muscle glycogen content; thus, the energy status of a muscle affects the release of IL-6 in response to exercise. The vast majority of studies also show that IL-6 is released on acute exercise but not in response to exercise training. An excellent overview and comparison of different published exercise studies on IL-6 is found in a review by Catoire and Kersten (2015). These authors concluded that the lack of induction of IL-6 by acute exercise is related to either the duration of exercise or the application of a moderate type of exercise. It appears that long-lasting, acute exercise sessions with concomitant depletion of glycogen stores are needed to increase plasma IL-6 levels, and it can be speculated that muscle damage contributes to the increase in IL-6. It was also reported that acute exercise leads to an increased expression of IL-6 in adipose tissue (Rosa Neto et al., 2009), making it likely that at least part of circulating IL-6 in response to acute exercise is not originating from skeletal muscle. Interestingly, exercise training reduces IL-6 gene expression and plasma levels, supporting the notion that IL-6 is not mediating the effects of moderate exercise training. Nevertheless, in vitro data clearly support the view that IL-6 may affect the metabolic performance of skeletal muscle cells. Both basal and insulin-stimulated glucose uptake was shown to be increased in myocytes treated with IL-6 due to translocation of the glucose transporter Glut4. These effects were shown to involve activation of AMPK (Al-Khalili et al., 2006). Evidence also exists that muscle-derived IL-6 exerts an antiinflammatory action. As pointed out by Karstoft and Pedersen (2016), the controversial results regarding the role of IL-6 in metabolic homeostasis may be related to its complex signaling biology. Thus, IL-6 may signal via the IL-6 receptor constitutively expressed on the cell surface, and this is thought not to generate a proinflammatory signal. Alternatively, when IL-6 binds to the soluble form of the IL-6 receptor, it binds to gp130, a cell-surface protein expressed in inflammatory cells. This complex scenario of IL-6 signaling points to a cell context–specific function of this cytokine and helps to understand why IL-6 may act as a positive regulator of metabolism in the context of exercise and, at the same time, play a key role in inflammatory processes.

Available data also suggest that IL-6 may play an important role in the muscle to pancreas crosstalk. This may take place in an indirect way in that exercise-induced IL-6 stimulates the release of glucagon-like peptide-1 from intestinal L-cells. This gut hormone is known to regulate insulin secretion and glycemia and to protect β-cell mass. In a very recent study, it was shown that exercise-induced IL-6 is also able to directly control β-cell viability (Paula et al., 2015). This study was based on earlier observations showing that exercise training is able to reduce β-cell damage in type 1 diabetic animals. In their work, Paula et al. (2015) now used wild-type and IL-6–knockout (KO) mice subjected to exercise. Islets from control and trained mice were subsequently exposed to inflammatory cytokines and apoptotic pathways were analyzed. It was demonstrated that exercise substantially reduces β-cell death and that this process is directly related to IL-6 signaling in the β-cells (Paula et al., 2015). It will be extremely important to assess in future studies whether these findings can be translated to the human setting, raising the possibility that exercise programs may prevent or delay the development of diabetes.

After the discovery of IL-6 as a secretory product of skeletal muscle, researchers started to investigate a potential myokine function of other members of the interleukin family. In 2010, Haugen et al. (2010) reported on the identification of IL-7 as a novel exercise-regulated myokine. As a member of the interleukin superfamily 2, IL-7 is involved in T- and B-cell maturation, but its biological role as a myokine is rather unexplored. It was found to be increased in plasma in young healthy men after acute resistance exercise and may play a role in muscle development (Kraemer et al., 2014). IL-7 may represent an exercise factor, but much more work is needed to allocate a functional role to this myokine. In contrast to IL-7, IL-13 was recently described as a myokine with an autocrine function that is able to augment glucose uptake and oxidation as well as glycogen synthesis in human myotubes (Jiang et al., 2013). Importantly, in type 2 diabetic patients, a substantially reduced serum level of IL-13 was observed in agreement with a reduced release of IL-13 from these subjects. Muscular IL-13 appears to be regulated by microRNA let-7, but data on exercise-dependent regulation of this myokine are not available. IL-8 and IL-15 also belong to the interleukin family 2 and are well-known proinflammatory cytokines. The impact of these two interleukins as an exercise factor is currently unclear, but available data suggest that only on very high-intensity exercise going along with muscle damage, plasma levels of IL-8 and IL-15 may increase (Catoire and Kersten, 2015). Specifically for IL-8, an increase in plasma levels was only found after an ultraendurance exercise bout (Marklund et al., 2013). IL-8 has been suggested to play a role in muscular angiogenesis, but this needs further investigation. Compared with IL-8, more data are available for IL-15 making it likely that this cytokine could play a role as an exercise factor both after acute and chronic conditions. A small but significant increase of circulating IL-15 was observed in untrained healthy young men after 30 min of treadmill running (Tamura et al., 2011). Interestingly, IL-15 appears to respond much more prominent to a resistance exercise program. After 8 weeks of moderate resistance exercise, the plasma level of IL-15 increased by 250%, whereas the plasma level of IL-6 was nearly unaltered (Yeo et al., 2012). Earlier data suggested that IL-15 may increase fatty acid oxidation in skeletal muscle, and it could therefore play a role in the regulation of body composition and adipose tissue mass (Quinn et al., 2009).

The signaling of IL-15 involves a specific IL-15 receptor alpha (IL-15 Rα) working in concert with the IL-2 receptor beta. Interestingly, disruption of the IL-15 Rα results in remodeling of skeletal muscle to a more oxidative phenotype along with increased levels of circulating IL-15. This potential autocrine effect of IL-15 was recently confirmed by in vitro studies using myocytes from IL-15R—KO mice and C2C12 cell cultures (O'Connell and Pistilli, 2015). It was shown that recombinant IL-15 increases transcriptional expression of the prooxidative genes PGC1-α and PPAR-δ and in myogenic differentiation of cells with a greater mitochondrial density. This very interesting property of IL-15 needs further investigation, as it may be useful for counteracting obesity development by increasing energy expenditure and fatty acid oxidation. This is a current and future hot topic of ongoing research, and other myokines potentially involved in this process will be discussed in the next

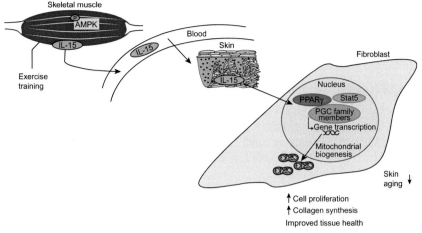

FIGURE 3.1

Graphical illustration of the effects of exercise-induced muscle interleukin-15 (IL-15) signaling to skin tissue.

Reproduced with permission from Crane, J.D., MacNeil, L.G., Lally, J.S., Ford, R.J., Bujak, A.L., Brar, I.K., Kemp, B.E., Raha, S., Steinberg, G.R., Tarnopolsky, M.A., 2015. Exercise-stimulated interleukin-15 is controlled by AMPK and regulates skin metabolism and aging. Aging Cell 14, 625–634.

section. A very interesting novel crosstalk function of IL-15 has been recently reported by Crane et al. (2015) by demonstrating that endurance exercise attenuates age-associated changes to skin in both humans and mice, involving IL-15 as a novel regulator of mitochondrial function in aging skin. These authors found a reduced epidermal thickness in exercising individuals compared with sedentary controls, and this effect could be correlated to circulating IL-15 levels by using a neutralizing antibody. Furthermore, by using muscle-specific AMPK-KO mice, it was shown that exercise controls IL-15 expression at least partly by this key regulator of metabolic function. This new muscle–skin crosstalk extends the myokine concept to an additional organ and emphasizes the complex scenario of the endocrine function of skeletal muscle (see Fig. 3.1).

NOVEL MYOKINES WITH POTENTIAL THERAPEUTIC APPLICATIONS

Given the beneficial effects of exercise training and the advanced understanding of the role of myokines in this process, many studies have now addressed the issue of contraction-regulated myokines as potential therapeutic targets. Specifically, the description of irisin as a novel myokine that was suggested to promote the so-called browning of white adipose tissue (Bostrom et al., 2012) has triggered a tremendous interest in exercise-regulated myokines. The reason is that irisin was claimed to increase thermogenesis and thus energy expenditure, a new way to combat obesity. Although irisin has remained highly controversial, some other myokines may in fact be of interest for future therapeutic approaches.

FOLLISTATIN-LIKE 1: A CARDIOPROTECTIVE MYOKINE

Members of the follistatin (FST) family act as binding partners for proteins of the TGF-β family such as myostatin. Follistatin-like 1 (FSTL1) is a secreted glycoprotein of 45−55 kDa belonging to this family of proteins, although it has only limited homology to FST. FSTL1 is definitely also an adipokine, and thus an adipomyokine, being highly expressed in preadipocytes and downregulated during differentiation (Raschke and Eckel, 2013). However, we discuss it here, as it is a paradigm for a potentially very interesting myokine with future impact as a therapeutic target. FSTL1 is secreted by human skeletal muscle cells, and plasma levels are increased after acute endurance exercise (Gorgens et al., 2013). Interestingly, proinflammatory cytokines were found to augment the release of FSTL1 from human myotubes. Because FSTL1 is thought to promote endothelial cell function and revascularization in ischemic tissues (Ouchi et al., 2008), the increased level of FSTL1 may counteract the deleterious effects of cytokines on the skeletal muscle vascular system. As pointed out before, FSTL1 expression and release is not limited to skeletal muscle. In addition to adipose tissue, it is also produced by endothelial cells and cardiomyocytes (Shimano et al., 2011). Expression of FSTL-1 in cardiac muscle is induced by ischemia, and FSTL-1 was shown to prevent cardiac injury by activating AMPK and inhibiting apoptosis and inflammation (Ogura et al., 2012). Because type 2 diabetes and the metabolic syndrome are associated with cardiovascular disease, it would be very important to assess the circulating level of this myokine and the effects of exercise in a diabetic cohort. Such data are presently not available, and therapeutic applications of FSTL1 need to be considered.

FIBROBLAST GROWTH FACTOR 21: A POTENTIAL INSULIN SENSITIZER

Fibroblast growth factor 21 (FGF21) together with FGF19 and FGF23 are endocrine members of the FGF superfamily. These hormonelike molecules lack the heparin-binding property and are released into the circulation exerting pleiotropic functions including bile acid synthesis (for FGF19), vitamin D metabolism (for FGF23), and the regulation of glucose and lipid metabolism by FGF21 (Jin et al., 2016). FGF21 is expressed in a number of tissues including the liver, adipose tissue, and others. FGF21 was first described as an Akt-regulated myokine, as its expression and secretion was found to be upregulated in C2C12 myocytes (Izumiya et al., 2008). Pharmacological studies have shown that FGF21 normalizes glucose and lipid homeostasis and, in that way, may prevent the development of obesity and diabetes. However, is FGF21 really an exercise regulated myokine? A stimulatory effect of acute exercise on FGF21 mRNA has been reported (Catoire et al., 2014b), but studies on the effect of exercise on the circulating level of FGF21 have remained controversial, at least in humans (Eckardt et al., 2014). It has been suggested that hepatic FGF21 as a major source of this molecule may contribute to increased serum levels after exercise involving exercise-induced lipolysis. Thus, the source of exercise-induced FGF21 needs further investigation, but the functional relevance of this protein merits specific attention. FGF21 acts on a variety of tissues by regulating carbohydrate and lipid metabolism, enhancing insulin sensitivity, and augmenting glucose uptake in skeletal muscle in both white and brown adipose tissue (BAT) (Jin et al., 2016). The downstream signaling of FGF21 remains poorly characterized, and it requires FGF receptors and the presence of β-Klotho, a protein being restricted to metabolic tissues. Substantial evidence supports the notion that adiponectin is a downstream effector FGF21. Thus, the beneficial effects of FGF21 are completely abolished in adiponectin KO mice (Lin et al., 2013). The complex scenario of FGF21 action is schematically presented in Fig. 3.2. It has

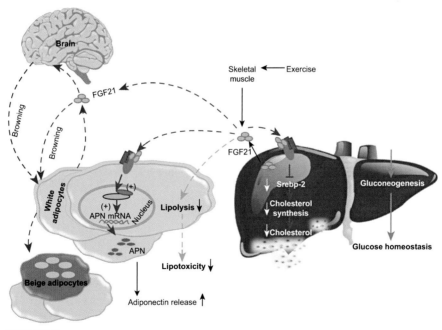

FIGURE 3.2

Fibroblast growth factor 21 (FGF21) exerts its cardiovascular protective activity via mediating the crosstalk between adipose tissue, brain, liver, and blood vessels. *APN*, adiponectin; *Srebp-2*, sterol regulatory element-binding protein 2.

Reproduced with permission from Jin, L., Lin, Z., Xu, A., 2016. Fibroblast growth factor 21 protects against atherosclerosis via fine-tuning the multiorgan crosstalk, Diabetes Metab. J. 40(1), 22–31. https://doi.org/10. 4093/dmj.2016.40.1.22. Copyright © 2016 Korean Diabetes Association

been suggested that FGF21 may act by fine-tuning the multiorgan crosstalk between the liver, brain, adipose tissue, and vasculature (Jin et al., 2016). Here, we add skeletal muscle to this multiorgan scenario, and physical exercise may positively modulate the FGF21 axis by multiple mechanisms. FGF21 has also been reported to promote the so called white-to-brown shift of adipocytes to a brite (brown-in-white)/beige phenotype, either directly or via the brain/adipocyte crosstalk (Douris et al., 2015). Increased energy expenditure by brite adipocytes may represent an additional feature of the beneficial effects of FGF21. A phase 1b clinical trial using the long-acting FGF21 analogue LY2405319 in obese, type 2 diabetic patients showed positive effects on body weight and insulin levels, along with a marked elevation in circulating adiponectin (Gaich et al., 2013). This raises great expectations, and it may be envisaged that with ongoing research novel multiorgan crosstalk regulators will be identified that may finally be used for new therapeutic approaches.

IRISIN: A CONTROVERSIAL MYOKINE

The well-accepted notion of the health-promoting effects of physical exercise and the myokine concept of organ communication both triggered an intensive search for novel myokines that could represent key mediators of this process. This issue gained additional interest by the discovery of BAT in humans in 2009 (Cypess et al., 2009; van Marken Lichtenbelt et al., 2009; Saito et al., 2009; Virtanen et al., 2009; Zingaretti et al., 2009) and the discovery of a third type of fat cells termed "brite" adipocytes (Petrovic et al., 2010). These cells share a number of biochemical properties and specifically the thermogenic potential with brown adipocytes, but they are derived from different precursor cells. These observations raised great expectations, and the possibility to direct the differentiation of adipose stem cells into brite adipocytes (a process termed as browning of white fat) has been extensively investigated in the last 5 years, most importantly, as an option to increase energy expenditure and to combat obesity and its complications.

During the search for potential regulators of fat cell browning, a novel hormone-like myokine termed irisin (named after the Greek messenger of the Gods, Iris) was reported to activate a white-to-brown transition in both rodents and humans (Bostrom et al., 2012). In this original report, irisin was claimed to be released after physical exercise and promoted as a new molecule that can mimic the positive effects of regular exercise on metabolic control and health ("sports in a pill"). Unfortunately, irisin turned out to be the most controversial myokine, and there is still a gap of knowledge regarding the precise function of this protein, specifically in humans (Elsen et al., 2014). This topic will be addressed in the next passage, when discussing the muscle–adipose tissue crosstalk. Before looking on these more functional properties, it is important to consider different aspects of the physiology of irisin, including its secretion and regulation by exercise.

Bostrom et al. (2012) studied a PGC1-α–overexpressing mouse model, as this transcriptional coactivator is well established as a key regulator of energy metabolism, mitochondrial function, and muscle gene expression. Besides several other genes, they observed a strong upregulation of *Fndc5* mRNA, a gene coding for fibronectin type III domain containing protein 5 (FNDC5). This transmembrane protein was first described in 2002 (Teufel et al., 2002), but its function remained largely unexplored with some evidence for a link to differentiation of myoblasts and neurones. The FNDC5 protein contains a signal peptide, fibronectin type III repeats, and a transmembrane domain. Bostrom et al. (2012) reported that the extracellular N-terminal part of FNDC5 is cleaved and released to the circulation. This 112-amino-acid protein contains the major part of the fibronectin III domain, was called irisin, and was described to be 100% identical in mice and humans. Until now, the cleavage process and the enzyme(s) mediating this step have remained elusive. Furthermore, irisin was considered as an exercise-regulated myokine based on the observation that the *Fndc5* gene is upregulated in mouse skeletal muscle after 3 weeks of freewheel running (Bostrom et al., 2012). Some key facts about irisin are summarized in Box 3.2.

Since the original description in 2012, numerous studies have analyzed the effect of acute and chronic exercise on the expression of FNDC5 and the release of irisin in both rodents and humans, with rather contradictory results, specifically in humans. A detailed analysis of different human exercise cohorts analyzed for *Fndc5* mRNA expression in skeletal muscle after different modes of exercise is presented by Elsen et al., (2014). Out of 15 human exercise studies (both acute and chronic exercise protocols), only four studies including the original Boström paper were able to monitor an increased expression of the *Fndc5* gene. However, not all exercise studies have measured induction of PGC1-α in parallel. Because there might be a tight correlation between this transcriptional coactivator and FNDC5/irisin, this is an important aspect. This issue has recently been addressed by Norheim et al. (2014). A cohort of sedentary men was submitted to a 12-week exercise intervention of combined endurance and strength training. Analysis of muscle biopsies indicated a slight increase of PGC1-α in parallel to a 1.4-fold increase of *Fndc5* mRNA, confirming the initial report. However, acute exercise did not affect expression of the *Fndc5* gene, although PGC1-α was substantially (7.4-fold) upregulated under these conditions. This lack of correlation between PGC1-α and FNDC5 was also confirmed by other investigators (Pekkala et al., 2013). Norheim et al. (2014) also used primary human myotubes to delineate a potential correlation between PGC1-α and FNDC5. Cells

BOX 3.2 KEY FACTS ABOUT IRISIN

- Irisin is a 12 kDa protein (unglycosylated) first described in 2012. It represents the extracellular, N-terminal part of the transmembrane protein FNDC5 and was suggested to be released from contracting muscle.
- Using untargeted mass spectrometry, irisin could be detected in the circulation of human subjects.
- The parent molecule FNDC5 is expressed in many tissues including the cardiac muscle, brain, and skin.
- The effect of exercise on FNDC5 expression and circulating irisin remains controversial.

were treated with exercise-mimicking agents such as caffeine, ionomycin, and forskolin. Although PGC1-α was substantially induced in response to these agents, the level of Fndc5 mRNA was even decreased under these conditions. As outlined before, EPS of human myotubes (see also Chapter 6) represents an excellent model to study muscle contraction under well-defined in vitro conditions. When using this model, it was found that similar to the work using exercise mimetics, *Fndc5* mRNA is not upregulated after EPS, despite an augmented level of PGC1-α in this model (Raschke et al., 2013b). Based on these data, it appears that at least in humans FNDC5 is not a downstream target of PGC1-α, and the potential regulation by exercise may involve secondary processes that yet have to be defined.

Irisin is not only present in skeletal muscle, but also in the cardiac muscle, brain, and skin and in smaller amounts in the liver, pancreas, and other tissues (Aydin et al., 2014). Thus, like many other molecules, irisin lacks specificity and cannot be considered only as a myokine. The contribution of all these tissues to circulating irisin has never been examined. Most importantly, the level of circulating irisin in different studies exhibits tremendous variation even among the same species. For humans, values between 0.01 and 2000 ng/mL have been reported using different ELISA assays. This raised considerable concern regarding the validity of these data and the initial description by Bostrom et al. (2012). The existence of human irisin has even been questioned for several reasons. First, the presence of irisin in human plasma was originally analyzed by Western blot using an antibody that detects the C-terminal region of the FNDC5 protein. As the N-terminal part is cleaved and released as irisin, the C-terminal domain of FNDC5 remains intracellular and should not be detected in the circulation. Second, and most importantly, it was demonstrated that the human *Fndc5* gene exhibits a mutation in the conserved start codon changing ATG to ATA (Raschke et al., 2013b). This was shown to result in a very low translation efficiency of human FNDC5, making it likely that in humans irisin is only circulating at a very low level. This was indeed confirmed recently in a very elegant study using targeted mass spectrometry (Jedrychowski et al., 2015). Using this approach, irisin was detected as a 12 kDa protein corresponding to the expected molecular mass of deglycosylated irisin. It was found at a level of about 0.2−0.3 nM, which is much lower than most reports in the literature using ELISA technologies. A very slight increase of only 20% was observed after 12 weeks of exercise using this technology. An interesting meta-analysis on the effects of chronic exercise on circulating irisin was recently applied to randomized controlled trials and nonrandomized studies. Importantly, chronic exercise training was found to decrease the circulating irisin level in the controlled studies, with inconsistent results in the nonrandomized studies (Qiu et al., 2015). Overall, the controversy about irisin as a myokine and its relation to exercise is still ongoing. Future research needs to determine the precise mechanisms of irisin release by skeletal muscle and the impact of exercise. Furthermore, the irisin receptor has not yet been discovered, and it would be instrumental in dissecting the signaling pathways of this molecule in myocytes and other cells. In addition to adipose tissue (see next passage), irisin may act on other cell types, and this needs to be explored in the future.

MUSCLE—ADIPOSE TISSUE CROSSTALK

As described in detail in Chapter 2, the adipocyte—myocyte axis can be considered as a paradigm of negative crosstalk between two tissues, and the impact of this communication for induction of muscle insulin resistance has been extensively investigated. However, on extension of the crosstalk scenario and with the development of the myokine concept, extensive investigation aimed to understand the bidirectional crosstalk between these tissues and specifically the role of exercise training on adipose tissue adaptation. As outlined before, the rediscovery of BAT in humans and the recognition that the so-called brite adipocytes play a key role for systemic energy metabolism (Cohen et al., 2014) have initiated a huge number of studies addressing the question whether exercise training (and hence potentially also myokines) may promote BAT activity and the browning of white fat. Although extensively studied, this issue has remained controversial, at least in humans. An excellent review on this topic has been published recently (Stanford and Goodyear, 2016).

Regular physical activity is well known to prevent an increase in body weight and lipid accumulation. Furthermore, a number of studies have shown that, independent of weight loss, regular exercise leads to a substantial remodeling of adipocytes with an increased number of small adipocytes containing less lipid and an increased expression of the glucose transporter GLUT4 and PGC1-α (Gollisch et al., 2009). Very strong evidence indicating that exercise-induced adaptations of adipose tissue exert beneficial effects on systemic metabolism was recently presented by Stanford et al. (2015). These authors reported that transplantation of subcutaneous fat from trained mice into the visceral cavity of sedentary mice induces improved glucose homeostasis in these mice. This effect could be attributed to a substantial change of white adipocytes to a brite phenotype (Stanford et al., 2015). This report stands in line with a number of other mouse studies that observed an increased abundance of brite adipocytes in subcutaneous adipose tissue after endurance exercise training (running, swimming). Although well documented in these rodent studies, the primary physiological function of exercise-induced browning has remained unclear, and this process appears hard to explain from a physiological point of view, for several reasons. First, contracting muscle already generates heat, and it appears not reasonable to induce a thermogenic process on top. Second, contracting muscle is an energy-consuming organ, and the induction of an additional energy-consuming site in the body appears not feasible. Finally, skeletal muscle relies on fatty acids supplied by white adipose tissue. Changing to a brite phenotype would limit the capacity to provide substrates to the working muscle; hence, physiologically an exercise-induced browning appears not meaningful (Kelly, 2012; Nedergaard and Cannon, 2014). Nevertheless, an ever increasing number of agents that are able to induce browning have been reported, including lactate, β-aminoisobutyric acid, BDNF, and several myokines such as meteorin-like 1 and irisin, which has already been described in detail before. Fig. 3.3 summarizes our current knowledge on the crosstalk between skeletal muscle and adipose tissue, mostly based on rodent studies.

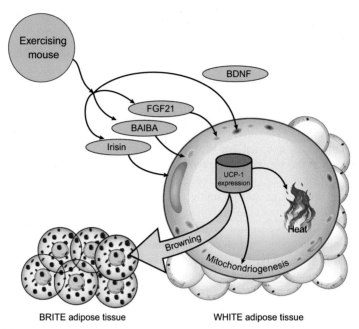

FIGURE 3.3

Crosstalk between skeletal muscle and adipose tissue.

In the human setting, the situation is much more complex, and the existing data suggest that exercise training has no effect on the brite phenotype in human adipose tissue. The group of van Marken Lichtenbelt reported no difference in brite cell markers when comparing subcutaneous white fat biopsies from sedentary subjects with that of endurance-trained subjects (Vosselman et al., 2015). Most importantly, in this study, it was also reported that regular endurance exercise training is associated with a decreased cold-induced activity of BAT. This goes in line with the physiological considerations that activation of a thermogenic tissue or process during exercise is not meaningful. Thus, the role of myokines in controlling adipose tissue function remains unsettled, and this is specifically the case for the novel myokine irisin.

In the first report on the browning function of irisin (Bostrom et al., 2012), the authors treated murine preadipocytes isolated from the inguinal fat depot with the parent irisin molecule FNDC5 during the differentiation process. It is important to note that subcutaneous fat, in contrast to other fat depots, exhibits the most prominent ability of responding to cold exposure with augmenting the brite phenotype (Walden et al., 2012). Bostrom et al. (2012) reported an increase in several brite marker genes and oxygen consumption. The latter is important, as it indicates a functional impact of irisin treatment. This was also confirmed by in vivo experiments using adenoviral-mediated overexpression of FNDC5 (Bostrom et al., 2012). These initial findings by the Spiegelman group were further elaborated, and in a second study, they reported that most likely only a subpopulation of subcutaneous white preadipocytes, selected by the cell surface marker CD137, is responsible to irisin-induced browning (Wu et al., 2012). They could show that CD137-sorted cells are highly responsive to FNDC5 and irisin in terms of upregulating the brite phenotype. Regarding the above-mentioned effect of exercise on browning of white fat in rodents and the corresponding data on the regulation of irisin in mice, it can be concluded that irisin is a novel myokine linking physical activity to an improved metabolic performance, mostly related to the white to brown transition observed in subcutaneous adipose tissue. It should also be mentioned that data suggesting a role for irisin as a regulator of glucose metabolism exist, involving adipose tissue but potentially also other targets such as skeletal muscle itself and the liver. However, all these data were generated in animal studies; human studies confirming these effects of irisin are currently lacking.

In fact, the translation of the irisin concept to the human setting has been hampered by several issues. First, as described above, in contrast to rodents, exercise training does not mediate browning of white fat in humans. Second, the existence of irisin in humans was doubted owing to a noncanonical ATA start codon. Third, very few data exist for analyzing the direct effect of irisin on browning of human adipocytes. In 2013, it was reported that neither irisin nor FNDC5 was able to induce browning in primary human preadipocytes obtained from subcutaneous fat of different donors (Raschke et al., 2013b). Even very high concentrations of irisin (up to 600 ng/mL) or experiments with cells highly expressing CD137 did not show any effect on the browning program, whereas BMP7, a well-established browning agent, induced a prominent effect under these conditions. This was confirmed in another study, when primary adipocytes were isolated from the subcutaneous or the omental depot (Lee et al., 2014). However, these authors found a strong induction of brown marker genes in adipocytes isolated from neck biopsies by treatment with FNDC5. These cells also exhibited an increased oxygen consumption after FNDC5 treatment and responded to β-adrenergic activation (Lee et al., 2014). Thus, it appears that the irisin response in humans may depend on the adipocyte lineage and the adipose tissue depot. Because the vast majority of human adipose depots are pure white, the percentage of cells that may undergo browning in response to irisin can be assumed to be rather small, with no or a very limited contribution to the metabolic performance. A very elegant study using laser-scanning cytometry to quantify ex vivo human adipocyte browning was published recently (Kristof et al., 2015). This technology allows measurement of brown adipocyte differentiation at a single cell level in a highly replicative manner. Using this approach, the authors tested the effect of irisin on browning of adipocytes isolated from human abdominal subcutaneous adipose tissue. They were able to show an increased expression of *Ucp1* and *Cidea* (two brite marker genes) and an increased oxygen consumption. Unfortunately, they only used a rather high concentration of irisin (250 ng/mL, 100-fold above the physiological level), and additional studies using this very interesting experimental tool need to be conducted in the future. Most recently, a paper appeared reporting that irisin exerts dual effects on both browning and adipogenesis of human white adipocytes (Zhang et al., 2016). In this study, irisin upregulated the expression of browning-associated genes in mature adipocytes and in fresh adipose tissue. Most importantly, irisin was found to inhibit adipogenic differentiation of cultured preadipocytes. However, irisin was found to promote osteogenic differentiation. Because reduced adipogenesis is thought to play an important role in hypertrophic obesity and the pathogenesis of insulin resistance (Gustafson et al., 2015), this issue needs further investigation.

To sum up, the muscle—adipose axis is a key player in mediating at least some of the beneficial effects of physical activity and regular exercise training. The muscle secretome is definitely involved in this process, but given the complex adaptation scenario in nearly every organ in response to exercise, it remains difficult to dissect these pathways and to allocate a key function to one or several myokines. Irisin started with great expectations as potentially being such a player, but its role in humans has been overestimated and remains ill-defined. Most likely, it is a combination of several myokines and metabolites together with additional pathways that helps to remodel adipose tissue and improve its metabolic performance.

THE MUSCLE—BONE CONNECTION

The close functional and developmental relationship between muscle and bone has been demonstrated in a number of clinical and experimental studies. The so-called "bone—muscle unit" concept is based on the lifelong association between lean muscle mass and the bone mineral content and reflects the tight functional association of these two tissues, which is mediating locomotion (Ferretti et al., 1998; Rauch et al., 2004). It is also well known that different physio- and pathophysiological stimuli exert a concomitant effect on both bone and muscle, supporting the notion that the musculoskeletal system is orchestrated by a multidirectional crosstalk system. At first hand, this involves the myokines, and bone is definitely a very important target for the secretory output of skeletal muscle (see a detailed description of this issue below). Secondly, some osteokines (such as IGF-1, OCN, PGE2, and others) produced by both osteoblasts and osteocytes were recently described and are thought to affect muscle development (Dallas et al., 2013). However, a much more complex picture of organ crosstalk arises, when additional tissues such as cartilage, adipose tissue, and tendons are integrated into a more comprehensive picture of the bone—muscle unit. This was elegantly summarized in a recent review by Tagliaferri et al. (2015) and is presented in Fig. 3.4. This scenario is a very impressive example of multidirectional crosstalk between different tissues involving a host of molecules released from these tissues. Some molecules such as myostatin may play an overarching role by inhibiting muscle, bone, and cartilage development, with a positive effect on adipose tissue. Adipose tissue inflammation is known to promote osteoporosis and sarcopenia, being at least partly related to the crosstalk scenario depicted in Fig. 3.4.

Zooming in on skeletal muscle and the role of exercise, it is now well accepted that bone formation is regulated by both mechanical loading and the paracrine effect

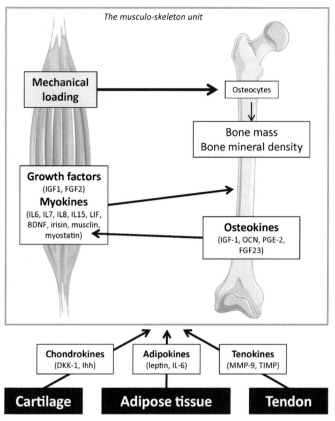

FIGURE 3.4

The muscle—bone unit.

Reproduced with permission from Tagliaferri, C., Wittrant, Y., Davicco, M.J., Walrand, S., Coxam, V., May 2015. Muscle and bone, two interconnected tissues. Ageing Res. Rev. 21, 55–70. https://doi.org/10.1016/j.arr.2015. 03.002.

of secreted myokines (Hamrick et al., 2010). The growth factor IGF-1 is such a myokine with a well-documented function for bone metabolism. Thus, it was shown that overexpression of IGF-1 in muscle results in an increased muscle mass and a concomitant enhancement of bone parameters (Banu et al., 2003). Regarding the classical myokines described before, good evidence exists for a prominent role of IL-6 in the muscle—bone crosstalk (Hamrick, 2011). This was also confirmed by recent clinical investigations. The myokines IL-8, IL-15, LIF, and BDNF are also suggested to be involved in this process, but more future studies will be required to delineate the precise function of these molecules (Tagliaferri et al., 2015). Finally, an important negative regulator of muscle development is myostatin, a member of the TGF-β superfamily, also known as GDF-8. Myostatin may affect the osteogenic differentiation of mesenchymal stem cells (Elkasrawy et al., 2012), but until now, it is not clear if myostatin exerts direct effects on either osteoblasts or osteoclasts. Exciting findings were reported recently in that the myokine osteoactivin was found to induce transdifferentiation of myoblasts into osteoblasts, suggesting a role of the muscle secretome for controlling bone remodeling (Sondag et al., 2014). Future research in this area may help to discover new targets with potential impact for prevention of osteoporosis or other bone disorders.

CONCLUSION AND FUTURE DIRECTIONS

Physical activity and regular exercise training can be considered as an integral element of a healthy lifestyle, and personalized exercise programs contribute to the therapeutic portfolio for a number of apparently unrelated diseases. With the myokine concept, the endocrine function of skeletal muscle gained evidence, triggering a huge interest in the identification of key myokines with the potential impact of being new drug targets. At present, most interest is focused on the muscle—adipose tissue axis, specifically fired by the idea that certain myokines such as irisin may promote the brite, energy-expending adipocyte phenotype. This issue will definitely remain of great interest, and future work should address additional molecules such as IL-15 and several others. A gap of knowledge still exists regarding the regulation of the muscle secretome. Although contraction plays an important role, a deeper understanding of the regulatory signals, specifically in the context of disease related scenarios, such as inflammation, sarcopenia, cachexia, and many others, will be mandatory for further developing the myokine concept. This also applies to the question of individual myokine signatures. This issue is still completely unexplored, and extended knowledge in this field will be of considerable impact for future projects related to therapeutic applications of the myokine concept. Finally, we need to keep in mind that presently only about 5% of the myokinome is allocated to a specific function. Thus, there is still a tremendous amount of work to do, and bioinformatic approaches need to be applied to the analysis of a multicomponent communication network between skeletal muscle and other organs. The pancreas, skin, and bone are some examples discussed in this chapter, but beyond these tissues, a much more complex network may exist that needs to be discovered in the future.

REFERENCES

Al-Khalili, L., Bouzakri, K., Glund, S., Lonnqvist, F., Koistinen, H.A., Krook, A., 2006. Signaling specificity of interleukin-6 action on glucose and lipid metabolism in skeletal muscle. Mol. Endocrinol. 20, 3364–3375.

Aydin, S., Kuloglu, T., Aydin, S., Kalayci, M., Yilmaz, M., Cakmak, T., Albayrak, S., Gungor, S., Colakoglu, N., Ozercan, I.H., 2014. A comprehensive immunohistochemical examination of the distribution of the fat-burning protein irisin in biological tissues. Peptides 61, 130–136.

Banu, J., Wang, L., Kalu, D.N., 2003. Effects of increased muscle mass on bone in male mice overexpressing IGF-I in skeletal muscles. Calcif. Tissue Int. 73, 196—201.

Besse-Patin, A., Montastier, E., Vinel, C., Castan-Laurell, I., Louche, K., Dray, C., Daviaud, D., Mir, L., Marques, M.A., Thalamas, C., Valet, P., Langin, D., Moro, C., Viguerie, N., 2014. Effect of endurance training on skeletal muscle myokine expression in obese men: identification of apelin as a novel myokine. Int. J. Obes. 38, 707—713.

Bostrom, P., Wu, J., Jedrychowski, M.P., Korde, A., Ye, L., Lo, J.C., Rasbach, K.A., Bostrom, E.A., Choi, J.H., Long, J.Z., Kajimura, S., Zingaretti, M.C., Vind, B.F., Tu, H., Cinti, S., Hojlund, K., Gygi, S.P., Spiegelman, B.M., 2012. A PGC1-alpha-dependent myokine that drives brown-fat-like development of white fat and thermogenesis. Nature 481, 463—468.

Catoire, M., Alex, S., Paraskevopulos, N., Mattijssen, F., Evers-van Gogh, I., Schaart, G., Jeppesen, J., Kneppers, A., Mensink, M., Voshol, P.J., Olivecrona, G., Tan, N.S., Hesselink, M.K., Berbee, J.F., Rensen, P.C., Kalkhoven, E., Schrauwen, P., Kersten, S., 2014a. Fatty acid-inducible ANGPTL4 governs lipid metabolic response to exercise. Proc. Natl. Acad. Sci. U. S. A. 111, E1043—E1052.

Catoire, M., Kersten, S., 2015. The search for exercise factors in humans. FASEB. J. 29, 1615—1628.

Catoire, M., Mensink, M., Kalkhoven, E., Schrauwen, P., Kersten, S., 2014b. Identification of human exercise-induced myokines using secretome analysis. Physiol. Genom. 46, 256—267.

Cohen, P., Levy, J.D., Zhang, Y., Frontini, A., Kolodin, D.P., Svensson, K.J., Lo, J.C., Zeng, X., Ye, L., Khandekar, M.J., Wu, J., Gunawardana, S.C., Banks, A.S., Camporez, J.P., Jurczak, M.J., Kajimura, S., Piston, D.W., Mathis, D., Cinti, S., Shulman, G.I., Seale, P., Spiegelman, B.M., 2014. Ablation of PRDM16 and beige adipose causes metabolic dysfunction and a subcutaneous to visceral fat switch. Cell 156, 304—316.

Crane, J.D., MacNeil, L.G., Lally, J.S., Ford, R.J., Bujak, A.L., Brar, I.K., Kemp, B.E., Raha, S., Steinberg, G.R., Tarnopolsky, M.A., 2015. Exercise-stimulated interleukin-15 is controlled by AMPK and regulates skin metabolism and aging. Aging Cell 14, 625—634.

Cypess, A.M., Lehman, S., Williams, G., Tal, I., Rodman, D., Goldfine, A.B., Kuo, F.C., Palmer, E.L., Tseng, Y.H., Doria, A., Kolodny, G.M., Kahn, C.R., 2009. Identification and importance of brown adipose tissue in adult humans. N. Engl. J. Med. 360, 1509—1517.

Dallas, S.L., Prideaux, M., Bonewald, L.F., 2013. The osteocyte: an endocrine cell... and more. Endocr. Rev. 34, 658—690.

Deshmukh, A.S., Cox, J., Jensen, L.J., Meissner, F., Mann, M., 2015. Secretome analysis of lipid-induced insulin resistance in skeletal muscle cells by a combined experimental and bioinformatics workflow. J. Proteome Res. 14, 4885—4895.

Douris, N., Stevanovic, D.M., Fisher, F.M., Cisu, T.I., Chee, M.J., Nguyen, N.L., Zarebidaki, E., Adams, A.C., Kharitonenkov, A., Flier, J.S., Bartness, T.J., Maratos-Flier, E., 2015. Central fibroblast growth factor 21 Browns white fat via sympathetic action in male mice. Endocrinology 156, 2470—2481.

Eckardt, K., Gorgens, S.W., Raschke, S., Eckel, J., 2014. Myokines in insulin resistance and type 2 diabetes. Diabetologia 57, 1087—1099.

Elkasrawy, M., Immel, D., Wen, X., Liu, X., Liang, L.F., Hamrick, M.W., 2012. Immunolocalization of myostatin (GDF-8) following musculoskeletal injury and the effects of exogenous myostatin on muscle and bone healing. J. Histochem. Cytochem. 60, 22—30.

Elsen, M., Raschke, S., Eckel, J., 2014. Browning of white fat: does irisin play a role in humans? J. Endocrinol. 222, R25–R38.

Ferretti, J.L., Capozza, R.F., Cointry, G.R., Garcia, S.L., Plotkin, H., Alvarez Filgueira, M.L., Zanchetta, J.R., 1998. Gender-related differences in the relationship between densitometric values of whole-body bone mineral content and lean body mass in humans between 2 and 87 years of age. Bone 22, 683–690.

Gaich, G., Chien, J.Y., Fu, H., Glass, L.C., Deeg, M.A., Holland, W.L., Kharitonenkov, A., Bumol, T., Schilske, H.K., Moller, D.E., 2013. The effects of LY2405319, an FGF21 analog, in obese human subjects with type 2 diabetes. Cell Metabol. 18, 333–340.

Goldstein, M.S., 1961. Humoral nature of the hypoglycemic factor of muscular work. Diabetes 10, 232–234.

Gollisch, K.S., Brandauer, J., Jessen, N., Toyoda, T., Nayer, A., Hirshman, M.F., Goodyear, L.J., 2009. Effects of exercise training on subcutaneous and visceral adipose tissue in normal- and high-fat diet-fed rats. Am. J. Physiol. Endocrinol. Metab. 297, E495–E504.

Gorgens, S.W., Eckardt, K., Jensen, J., Drevon, C.A., Eckel, J., 2015. Exercise and regulation of adipokine and myokine production. Prog Mol Biol Transl Sci 135, 313–336.

Gorgens, S.W., Raschke, S., Holven, K.B., Jensen, J., Eckardt, K., Eckel, J., 2013. Regulation of follistatin-like protein 1 expression and secretion in primary human skeletal muscle cells. Arch. Physiol. Biochem. 119, 75–80.

Gustafson, B., Hedjazifar, S., Gogg, S., Hammarstedt, A., Smith, U., 2015. Insulin resistance and impaired adipogenesis. Trends Endocrinol. Metabol. 26, 193–200.

Hamrick, M.W., 2011. A role for myokines in muscle-bone interactions. Exerc. Sport Sci. Rev. 39, 43–47.

Hamrick, M.W., McNeil, P.L., Patterson, S.L., 2010. Role of muscle-derived growth factors in bone formation. J. Musculoskelet. Neuronal Interact. 10, 64–70.

Hartwig, S., Raschke, S., Knebel, B., Scheler, M., Irmler, M., Passlack, W., Muller, S., Hanisch, F.G., Franz, T., Li, X., Dicken, H.D., Eckardt, K., Beckers, J., de Angelis, M.H., Weigert, C., Haring, H.U., Al-Hasani, H., Ouwens, D.M., Eckel, J., Kotzka, J., Lehr, S., 2014. Secretome profiling of primary human skeletal muscle cells. Biochim. Biophys. Acta 1844, 1011–1017.

Haugen, F., Norheim, F., Lian, H., Wensaas, A.J., Dueland, S., Berg, O., Funderud, A., Skalhegg, B.S., Raastad, T., Drevon, C.A., 2010. IL-7 is expressed and secreted by human skeletal muscle cells. Am. J. Physiol. Cell Physiol. 298, C807–C816.

Izumiya, Y., Bina, H.A., Ouchi, N., Akasaki, Y., Kharitonenkov, A., Walsh, K., 2008. FGF21 is an Akt-regulated myokine. FEBS Lett. 582, 3805–3810.

Jedrychowski, M.P., Wrann, C.D., Paulo, J.A., Gerber, K.K., Szpyt, J., Robinson, M.M., Nair, K.S., Gygi, S.P., Spiegelman, B.M., 2015. Detection and quantitation of circulating human irisin by tandem mass spectrometry. Cell Metabol. 22, 734–740.

Jiang, L.Q., Franck, N., Egan, B., Sjogren, R.J., Katayama, M., Duque-Guimaraes, D., Arner, P., Zierath, J.R., Krook, A., 2013. Autocrine role of interleukin-13 on skeletal muscle glucose metabolism in type 2 diabetic patients involves microRNA let-7. Am. J. Physiol. Endocrinol. Metab. 305, E1359–E1366.

Jin, L., Lin, Z., Xu, A., 2016. Fibroblast growth factor 21 protects against atherosclerosis via fine-tuning the multiorgan crosstalk. Diabetes Metab. J 40, 22–31.

Karstoft, K., Pedersen, B.K., 2016. Skeletal muscle as a gene regulatory endocrine organ. Curr. Opin. Clin. Nutr. Metab. Care 19, 270–275.

Kelly, D.P., 2012. Medicine. Irisin, light my fire. Science 336, 42–43.

Kraemer, W.J., Hatfield, D.L., Comstock, B.A., Fragala, M.S., Davitt, P.M., Cortis, C., Wilson, J.M., Lee, E.C., Newton, R.U., Dunn-Lewis, C., Hakkinen, K., Szivak, T.K., Hooper, D.R., Flanagan, S.D., Looney, D.P., White, M.T., Volek, J.S., Maresh, C.M., 2014. Influence of HMB supplementation and resistance training on cytokine responses to resistance exercise. J. Am. Coll. Nutr. 33, 247−255.

Kristof, E., Doan-Xuan, Q.M., Bai, P., Bacso, Z., Fesus, L., 2015. Laser-scanning cytometry can quantify human adipocyte browning and proves effectiveness of irisin. Sci. Rep. 5, 12540.

Lambernd, S., Taube, A., Schober, A., Platzbecker, B., Gorgens, S.W., Schlich, R., Jeruschke, K., Weiss, J., Eckardt, K., Eckel, J., 2012. Contractile activity of human skeletal muscle cells prevents insulin resistance by inhibiting pro-inflammatory signalling pathways. Diabetologia 55, 1128−1139.

Lee, P., Linderman, J.D., Smith, S., Brychta, R.J., Wang, J., Idelson, C., Perron, R.M., Werner, C.D., Phan, G.Q., Kammula, U.S., Kebebew, E., Pacak, K., Chen, K.Y., Celi, F.S., 2014. Irisin and FGF21 are cold-induced endocrine activators of brown fat function in humans. Cell Metabol. 19, 302−309.

Lin, Z., Tian, H., Lam, K.S., Lin, S., Hoo, R.C., Konishi, M., Itoh, N., Wang, Y., Bornstein, S.R., Xu, A., Li, X., 2013. Adiponectin mediates the metabolic effects of FGF21 on glucose homeostasis and insulin sensitivity in mice. Cell Metabol. 17, 779−789.

Marklund, P., Mattsson, C.M., Wahlin-Larsson, B., Ponsot, E., Lindvall, B., Lindvall, L., Ekblom, B., Kadi, F., 2013. Extensive inflammatory cell infiltration in human skeletal muscle in response to an ultraendurance exercise bout in experienced athletes. J. Appl. Physiol. 114, 66−72.

Nedachi, T., Fujita, H., Kanzaki, M., 2008. Contractile C2C12 myotube model for studying exercise-inducible responses in skeletal muscle. Am. J. Physiol. Endocrinol. Metab. 295, E1191−E1204.

Nedergaard, J., Cannon, B., 2014. The browning of white adipose tissue: some burning issues. Cell Metabol. 20, 396−407.

Nikolic, N., Bakke, S.S., Kase, E.T., Rudberg, I., Flo Halle, I., Rustan, A.C., Thoresen, G.H., Aas, V., 2012. Electrical pulse stimulation of cultured human skeletal muscle cells as an in vitro model of exercise. PLoS One 7, e33203.

Norheim, F., Langleite, T.M., Hjorth, M., Holen, T., Kielland, A., Stadheim, H.K., Gulseth, H.L., Birkeland, K.I., Jensen, J., Drevon, C.A., 2014. The effects of acute and chronic exercise on PGC-1alpha, irisin and browning of subcutaneous adipose tissue in humans. FEBS J. 281, 739−749.

O'Connell, G.C., Pistilli, E.E., 2015. Interleukin-15 directly stimulates pro-oxidative gene expression in skeletal muscle in-vitro via a mechanism that requires interleukin-15 receptor alpha. Biochem. Biophys. Res. Commun. 458, 614−619.

Ogura, Y., Ouchi, N., Ohashi, K., Shibata, R., Kataoka, Y., Kambara, T., Kito, T., Maruyama, S., Yuasa, D., Matsuo, K., Enomoto, T., Uemura, Y., Miyabe, M., Ishii, M., Yamamoto, T., Shimizu, Y., Walsh, K., Murohara, T., 2012. Therapeutic impact of follistatin-like 1 on myocardial ischemic injury in preclinical models. Circulation 126, 1728−1738.

Ouchi, N., Oshima, Y., Ohashi, K., Higuchi, A., Ikegami, C., Izumiya, Y., Walsh, K., 2008. Follistatin-like 1, a secreted muscle protein, promotes endothelial cell function and revascularization in ischemic tissue through a nitric-oxide synthase-dependent mechanism. J. Biol. Chem. 283, 32802−32811.

Paula, F.M., Leite, N.C., Vanzela, E.C., Kurauti, M.A., Freitas-Dias, R., Carneiro, E.M., Boschero, A.C., Zoppi, C.C., 2015. Exercise increases pancreatic beta-cell viability in a model of type 1 diabetes through IL-6 signaling. FASEB. J. 29, 1805−1816.

Pedersen, B.K., 2013. Muscle as a secretory organ. Comp. Physiol. 3, 1337—1362.

Pedersen, B.K., Febbraio, M.A., 2012. Muscles, exercise and obesity: skeletal muscle as a secretory organ. Nat. Rev. Endocrinol. 8, 457—465.

Pedersen, B.K., Steensberg, A., Fischer, C., Keller, C., Keller, P., Plomgaard, P., Febbraio, M., Saltin, B., 2003. Searching for the exercise factor: is IL-6 a candidate? J. Muscle Res. Cell Motil. 24, 113—119.

Pedersen, B.K., Steensberg, A., Fischer, C., Keller, C., Keller, P., Plomgaard, P., Wolsk-Petersen, E., Febbraio, M., 2004. The metabolic role of IL-6 produced during exercise: is IL-6 an exercise factor? Proc. Nutr. Soc. 63, 263—267.

Pekkala, S., Wiklund, P.K., Hulmi, J.J., Ahtiainen, J.P., Horttanainen, M., Pollanen, E., Makela, K.A., Kainulainen, H., Hakkinen, K., Nyman, K., Alen, M., Herzig, K.H., Cheng, S., 2013. Are skeletal muscle FNDC5 gene expression and irisin release regulated by exercise and related to health? J. Physiol. 591, 5393—5400.

Petrovic, N., Walden, T.B., Shabalina, I.G., Timmons, J.A., Cannon, B., Nedergaard, J., 2010. Chronic peroxisome proliferator-activated receptor gamma (PPARgamma) activation of epididymally derived white adipocyte cultures reveals a population of thermogenically competent, UCP1-containing adipocytes molecularly distinct from classic brown adipocytes. J. Biol. Chem. 285, 7153—7164.

Qiu, S., Cai, X., Sun, Z., Schumann, U., Zugel, M., Steinacker, J.M., 2015. Chronic exercise training and circulating irisin in adults: a meta-analysis. Sports Med. 45, 1577—1588.

Quinn, L.S., Anderson, B.G., Strait-Bodey, L., Stroud, A.M., Argiles, J.M., 2009. Oversecretion of interleukin-15 from skeletal muscle reduces adiposity. Am. J. Physiol. Endocrinol. Metab. 296, E191—E202.

Raschke, S., Eckardt, K., Bjorklund Holven, K., Jensen, J., Eckel, J., 2013a. Identification and validation of novel contraction-regulated myokines released from primary human skeletal muscle cells. PLoS One 8, e62008.

Raschke, S., Eckel, J., 2013. Adipo-myokines: two sides of the same coin—mediators of inflammation and mediators of exercise. Mediat. Inflamm. 2013, 320724.

Raschke, S., Elsen, M., Gassenhuber, H., Sommerfeld, M., Schwahn, U., Brockmann, B., Jung, R., Wisloff, U., Tjonna, A.E., Raastad, T., Hallen, J., Norheim, F., Drevon, C.A., Romacho, T., Eckardt, K., Eckel, J., 2013b. Evidence against a beneficial effect of irisin in humans. PLoS One 8, e73680.

Rauch, F., Bailey, D.A., Baxter-Jones, A., Mirwald, R., Faulkner, R., 2004. The muscle-bone unit during the pubertal growth spurt. Bone 34, 771—775.

Rosa Neto, J.C., Lira, F.S., Oyama, L.M., Zanchi, N.E., Yamashita, A.S., Batista Jr., M.L., Oller do Nascimento, C.M., Seelaender, M., 2009. Exhaustive exercise causes an anti-inflammatory effect in skeletal muscle and a pro-inflammatory effect in adipose tissue in rats. Eur. J. Appl. Physiol. 106, 697—704.

Saito, M., Okamatsu-Ogura, Y., Matsushita, M., Watanabe, K., Yoneshiro, T., Nio-Kobayashi, J., Iwanaga, T., Miyagawa, M., Kameya, T., Nakada, K., Kawai, Y., Tsujisaki, M., 2009. High incidence of metabolically active brown adipose tissue in healthy adult humans: effects of cold exposure and adiposity. Diabetes 58, 1526—1531.

Shimano, M., Ouchi, N., Nakamura, K., van Wijk, B., Ohashi, K., Asaumi, Y., Higuchi, A., Pimentel, D.R., Sam, F., Murohara, T., van den Hoff, M.J., Walsh, K., 2011. Cardiac myocyte follistatin-like 1 functions to attenuate hypertrophy following pressure overload. Proc. Natl. Acad. Sci. U. S. A. 108, E899—E906.

Sondag, G.R., Salihoglu, S., Lababidi, S.L., Crowder, D.C., Moussa, F.M., Abdelmagid, S.M., Safadi, F.F., 2014. Osteoactivin induces transdifferentiation of C2C12 myoblasts into osteoblasts. J. Cell. Physiol. 229, 955–966.

Stanford, K.I., Goodyear, L.J., 2016. Exercise regulation of adipose tissue. Adipocyte 5, 153–162.

Stanford, K.I., Middelbeek, R.J., Townsend, K.L., Lee, M.Y., Takahashi, H., So, K., Hitchcox, K.M., Markan, K.R., Hellbach, K., Hirshman, M.F., Tseng, Y.H., Goodyear, L.J., 2015. A novel role for subcutaneous adipose tissue in exercise-induced improvements in glucose homeostasis. Diabetes 64, 2002–2014.

Steensberg, A., Keller, C., Starkie, R.L., Osada, T., Febbraio, M.A., Pedersen, B.K., 2002. IL-6 and TNF-alpha expression in, and release from, contracting human skeletal muscle. Am. J. Physiol. Endocrinol. Metab. 283, E1272–E1278.

Steensberg, A., van Hall, G., Osada, T., Sacchetti, M., Saltin, B., Klarlund Pedersen, B., 2000. Production of interleukin-6 in contracting human skeletal muscles can account for the exercise-induced increase in plasma interleukin-6. J. Physiol. 529 (Pt. 1), 237–242.

Tagliaferri, C., Wittrant, Y., Davicco, M.J., Walrand, S., Coxam, V., 2015. Muscle and bone, two interconnected tissues. Ageing Res. Rev. 21, 55–70.

Tamura, Y., Watanabe, K., Kantani, T., Hayashi, J., Ishida, N., Kaneki, M., 2011. Upregulation of circulating IL-15 by treadmill running in healthy individuals: is IL-15 an endocrine mediator of the beneficial effects of endurance exercise? Endocr. J. 58, 211–215.

Teufel, A., Malik, N., Mukhopadhyay, M., Westphal, H., 2002. Frcp1 and Frcp2, two novel fibronectin type III repeat containing genes. Gene 297, 79–83.

van Marken Lichtenbelt, W.D., Vanhommerig, J.W., Smulders, N.M., Drossaerts, J.M., Kemerink, G.J., Bouvy, N.D., Schrauwen, P., Teule, G.J., 2009. Cold-activated brown adipose tissue in healthy men. N. Engl. J. Med. 360, 1500–1508.

Virtanen, K.A., Lidell, M.E., Orava, J., Heglind, M., Westergren, R., Niemi, T., Taittonen, M., Laine, J., Savisto, N.J., Enerback, S., Nuutila, P., 2009. Functional brown adipose tissue in healthy adults. N. Engl. J. Med. 360, 1518–1525.

Vosselman, M.J., Hoeks, J., Brans, B., Pallubinsky, H., Nascimento, E.B., van der Lans, A.A., Broeders, E.P., Mottaghy, F.M., Schrauwen, P., van Marken Lichtenbelt, W.D., 2015. Low brown adipose tissue activity in endurance-trained compared with lean sedentary men. Int. J. Obes. 39, 1696–1702.

Walden, T.B., Hansen, I.R., Timmons, J.A., Cannon, B., Nedergaard, J., 2012. Recruited vs. nonrecruited molecular signatures of brown, "brite," and white adipose tissues. Am. J. Physiol. Endocrinol. Metab. 302, E19–E31.

Wu, J., Bostrom, P., Sparks, L.M., Ye, L., Choi, J.H., Giang, A.H., Khandekar, M., Virtanen, K.A., Nuutila, P., Schaart, G., Huang, K., Tu, H., van Marken Lichtenbelt, W.D., Hoeks, J., Enerback, S., Schrauwen, P., Spiegelman, B.M., 2012. Beige adipocytes are a distinct type of thermogenic fat cell in mouse and human. Cell 150, 366–376.

Yeo, N.H., Woo, J., Shin, K.O., Park, J.Y., Kang, S., 2012. The effects of different exercise intensity on myokine and angiogenesis factors. J. Sports Med. Phys. Fit. 52, 448–454.

Yoon, J.H., Kim, J., Song, P., Lee, T.G., Suh, P.G., Ryu, S.H., 2012. Secretomics for skeletal muscle cells: a discovery of novel regulators? Adv. Biol. Regul. 52, 340–350.

Zhang, Y., Xie, C., Wang, H., Foss, R.M., Clare, M., George, E.V., Li, S., Katz, A., Cheng, H., Ding, Y., Tang, D., Reeves, W.H., Yang, L.J., 2016. Irisin exerts dual effects on browning and adipogenesis of human white adipocytes. Am. J. Physiol. Endocrinol. Metab. 311, E530−E541.

Zingaretti, M.C., Crosta, F., Vitali, A., Guerrieri, M., Frontini, A., Cannon, B., Nedergaard, J., Cinti, S., 2009. The presence of UCP1 demonstrates that metabolically active adipose tissue in the neck of adult humans truly represents brown adipose tissue. FASEB. J. 23, 3113−3120.

Adipomyokines: An Extended View on the Crosstalk Scenario

CHAPTER OUTLINE

In the Chapter 3, we have described the myokine concept and the contribution of skeletal muscle to the organ crosstalk scenario due to its ability to secrete regulatory peptides and proteins in conjunction with physical exercise. As pointed out before, the skeletal muscle secretome turned out to be as complex as the adipokine profile, and even more important, a substantial overlap between the secretome of muscle and that of fat was observed (Giudice and Taylor, 2017; Carson, 2017; Petriz et al., 2017; Bomb et al., 2016). A paradigm of this overlap is the cytokine IL-6, a proinflammatory protein released by immune cells and in high amounts by enlarged and inflamed adipose tissue. On the other hand, IL-6 became the gold standard of a myokine and is definitely a positive regulator of metabolic homeostasis upon exercise. Excellent reviews on the multiple and pleiotropic functions of IL-6 should be considered for an additional reading (Rose-John et al., 2017; Hennigar et al., 2017; Sahibzada et al., 2017; Jordan et al., 2017). A fascinating and still principally unresolved issue of the crosstalk concept is the question, how it is possible that the same molecule would exert completely different, both beneficial and harmful, effects at the same time. In a publication from 2011 (Trayhurn et al., 2011), we used the term "adipomyokine" to describe this complexity and to highlight the existence of this overlap between the two tissues. It is now clear that exercise also affects the adipose tissue secretome, also contributing to the circulating level of adipomyokines and representing an additional aspect of physical exercise. In this chapter, we aim to provide a deeper look on the exercise-induced production and function of both adipokines and myokines, with a specific focus on several adipomyokines and their impact for metabolic regulation. The topic of adipomyokines

The Cellular Secretome and Organ Crosstalk. https://doi.org/10.1016/B978-0-12-809518-8.00004-0

has also gained considerable interest in the field of bone metabolism, and we will address this important aspect at the end of this chapter.

As pointed out before, the most extensively studied myokine is definitely IL-6, which is profoundly upregulated during acute exercise and assumed to play a key role in the antiinflammatory effect of acute exercise (Pedersen and Febbraio, 2005, 2008, 2012; Pedersen et al., 2003a,b, 2004; Febbraio and Pedersen, 2002, 2005). However, as described in Chapter 3, it has become evident that skeletal muscle secretes hundreds of myokines, with many of them being regulated by muscle contraction (Giudice and Taylor, 2017; Carson, 2017). At present, for most of these molecules, the biological function has remained elusive. An additional complexity results from the variety of physical activity in lifestyle intervention studies and substantial differences in the exercise training programs. The term acute exercise normally relates to a structured program performed under controlled intensity and duration. Such acute exercise bouts do not lead to major adaptations in muscle and other organs, but it is well known that insulin sensitivity can increase in the post-exercise phase between 24 and 48 h. Furthermore, training must be divided into endurance and strength training with multiple combinations of intensity, duration, and volume. Acute endurance exercise under controlled conditions is normally performed at about 70% of maximal oxygen consumption for a period of 30−60 min (Amiri Arimi et al., 2017; Myers et al., 2017).

The adipomyokine concept is presented in Fig. 4.1. It should be noted that in this scheme adipomyokines are molecules that are released by both skeletal muscle and

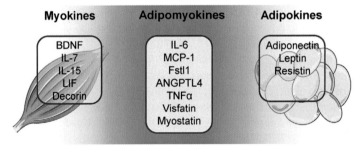

FIGURE 4.1

The adipomyokine concept. A search of original articles in PubMed was performed for the major exercise-regulated myokines and adipokines to identify molecules that were produced and secreted in both tissues. The term adipomyokines was used for proteins fulfilling both of these criteria. The search terms we used were "skeletal muscle" or "adipose tissue," "myokine" or "adipokine," and "exercise." *ANGPTL4*, angiopoietin-like protein 4; *BDNF*, brain-derived neurotrophic factor; *Fstl1*, follistatin-like 1; *LIF*, leukemia inhibitory factor; *MCP-1*, monocyte chemoattractant protein 1.

Reproduced with permission from Goergens, S., et al., 2015. Exercise and regulation of adipokine and myokine production. In: Claude Bouchard, editor, Progress in Molecular Biology and Translational Science. vol. 135. Academic Press, Burlington, pp. 313−336.

adipose tissue in response to exercise. These molecules are mediators of exercise and mediators of inflammation, most likely exerting a yet incompletely understood time- and concentration-dependent function. To gain a more comprehensive view on the molecular pathways involved in physical activity, an analysis of myokines, adipo-kines, and adipomyokines including adipose tissue is required. In fact, exercise training has been reported to reduce central adiposity independent of overall weight changes (Lee et al., 2005), and this may represent an additional mechanism of the antiinflammatory action of acute exercise mentioned earlier. Unfortunately, only very limited information is available regarding direct effects of exercise on adipo-kine and adipomyokine production.

By using the adipomyokine concept and including both skeletal muscle and adipose tissue, a more global view on the beneficial effects of different exercise programs and the underlying pathways can be obtained. In the following sections, we only considered human studies, and the molecules should be detectable in the circulation. This does not exclude autocrine/paracrine effects of the considered adipomyokines but emphasizes our view on the crosstalk between fat and muscle and the potential mediators of this process.

SKELETAL MUSCLE AS THE SOURCE OF MYOKINES AND ADIPOMYOKINES AFTER ACUTE AND CHRONIC EXERCISE

As already mentioned, IL-6 is the prototype of an adipomyokine, meaning it is expressed and released by both skeletal muscle and adipose tissue. It acts in an auto-crine/paracrine fashion within skeletal muscle and endocrine in a hormonelike fashion to mediate metabolic and antiinflammatory effects. IL-6 has been shown to trigger an antiinflammatory cascade by promoting the induction of antiinflammatory factors, such as IL-10 and IL-1ra (IL-1 receptor agonist), and to inhibit the production of the proinflammatory cytokines IL-1β and TNFα (Eckardt et al., 2014). Moreover, IL-6 plays a role in hypertrophic muscle growth (Serrano et al., 2008). Skeletal muscle secretes IL-6 into the circulation in response to acute muscle contractions. The increase of circulating IL-6 in response to acute exercise occurs in an exponential manner, and the maximum is reached at the end of the exercise session. The magni-tude by which IL-6 levels increase is related to the type of exercise, its duration, intensity, and the amount of muscle mass engaged in the exercise but is not affected by muscle damage. A very prominent increase of plasma IL-6 can be observed upon running, which involves several large muscle groups (Fischer, 2006). Several epidemi-ological studies suggest a negative association between the extent of regular physical activity and the basal plasma IL-6 levels; basal plasma IL-6 concentration is more associated with physical inactivity than other cytokines associated with the metabolic syndrome (Fischer, 2006). In contrast to acute exercise, data on the effect of regular exercise on circulating IL-6 concentrations are more controversial and the role of IL-6 under these conditions has been debated. Aerobic training for 10 months with over-weight elderly subjects (age $\geq$ 64 years) reduced basal plasma levels of IL-6, whereas strength training had no effect (Kohut et al., 2006). Another study with severely obese humans, which used a combination of hypocaloric diet and regular physical activity for 15 weeks, reported decreased basal plasma levels of IL-6 and reduced IL-6 mRNA expression in skeletal muscle with a parallel reduction of body weight by 13% (Bruun et al., 2006). However, some studies observed no changes in circulating IL-6 concentrations in response to chronic exercise (Rokling-Andersen et al., 2007). Additional aspects of IL-6 regulation by exercise have already been described in Chapter 3. Overall, IL-6 is released from contracting human skeletal muscle, and exercise-induced IL-6 has metabolic effects in humans, by affecting insulin-stimulated glucose disposal and fatty acid oxidation. In the basic inactive condition, the adipose tissue probably is the main source of IL-6, potentially with other metabolic effects than under physical activity. Two other members of the interleukin family, namely IL-7 and IL-15, have also been described as exercise factors and are presented in detail in Chapter 3.

Another member of the IL-6 superfamily has been classified as a myokine, namely leukemia inhibitory factor (LIF). LIF exerts multiple biological functions such as induction of satellite cell proliferation, an essential process for muscle hypertrophy and regeneration after muscle damage (Broholm et al., 2011). These authors reported in young healthy volunteers that LIF mRNA expression in skeletal muscle increased after acute aerobic exercise (4.5-fold) and heavy resistance exercise (9-fold). However, LIF protein level in skeletal muscle tissue was not changed, and LIF could not be detected in plasma in this study (Broholm et al., 2011). Owing to the lack of data in the literature, future studies should address the regulation of plasma LIF levels during different exercise protocols to determine whether the increased LIF mRNA expression in the muscle is reflected in the circulation.

Myostatin is a member of the transforming growth factor beta (TGF-β) superfamily that negatively regulates skeletal muscle size and was the first described myokine (Fan et al., 2017; Deng et al., 2017; Chen and Lee, 2016; Walker et al., 2016). When subjected to a moderate aerobic exercise training program for 6 months, overweight or obese men displayed a reduction of skeletal muscle expression and circulating levels of myostatin (Hittel et al., 2010). Moreover, acute endurance (Harber et al., 2009) or resistance exercise (Kim et al., 2005) was shown to attenuate myostatin mRNA expression in skeletal muscle.

Myostatin function can be antagonized by different proteins such as follistatin (FST) and decorin. Another example of an antagonistic protein of the TGF-β superfamily is FST-like 1 (Fstl1), which is a secreted glycoprotein belonging to the FST family of proteins (Shi and Massague, 2003). As pointed out before in Chapter 3, Fstl1 may have beneficial effects on ischemia—reperfusion injury in muscle and heart tissue associated with antiapoptosis (Ogura et al., 2012). We have recently shown that Fstl1 is a myokine expressed and secreted by primary human skeletal muscle cells. Moreover, Fstl1 plasma levels were increased after acute endurance exercise performed by young healthy men (Gorgens et al., 2013). It can be concluded that members of the TGF-β superfamily and their specific antagonists are strongly regulated by acute and chronic exercise to play a role in exercise-related restructuring processes of skeletal muscle.

Brain-derived neurotrophic factor (BDNF) has been described as an exercise-induced myokine (Matthews et al., 2009), although the protein and its receptor are most abundantly expressed in the brain. Acute endurance exercise increases plasma BDNF levels, most prominently upon high-intensity exercise (Schmidt-Kassow et al., 2012). In addition, Matthews et al. (2009) found significantly increased BDNF mRNA and protein abundance in skeletal muscle after acute endurance exercise. It was also reported that circulating BDNF concentrations increased in response to chronic endurance training (Griffin et al., 2011). However, a study using resistance exercise could not find an effect on BDNF plasma levels after acute or chronic exercise (Goekint et al., 2010). In humans, 70%−80% of plasma BDNF originates from the brain during both rest and after exercise, suggesting the brain as the major source of this factor (Rasmussen et al., 2009). It might be that muscle-derived BDNF acts primarily within skeletal muscle tissue, e.g., inducing lipid oxidation via AMPK activation, whereas brain-derived BDNF may act more systemically and may play a role in beneficial effects of exercise with regard to Alzheimer disease, depression, or impaired cognitive function (Mattson, 2012).

Angiopoietin-like protein 4 (ANGPTL4) is detected both in skeletal muscle and in adipose tissue and is therefore classified as adipomyokine that, in addition, is regulated by exercise (Kersten et al., 2009). Importantly, the induction of plasma ANGPTL4 after exercise may depend on the increase of free fatty acids. In healthy, untrained male volunteers, circulating ANGPTL4 was found to be increased after an acute endurance exercise program (Kersten et al., 2009). These authors also observed significantly increased plasma ANGPTL4 levels in healthy men after one-legged cycling exercise (Catoire et al., 2014). Interestingly, ANGPTL4 mRNA is more highly induced in the nonexercising leg than in the exercising leg. However, plasma ANGPTL4 levels were not changed after 2 weeks of intense endurance exercise training, or by 12 weeks endurance training. In agreement with these findings, Norheim et al. reported that acute endurance exercise significantly increased plasma ANGPTL4 levels and skeletal muscle ANGPTL4 mRNA. However, a chronic exercise intervention combining strength and endurance training for 12 weeks had no effect on basal plasma ANGPTL4 concentrations and skeletal muscle ANGPTL4 mRNA content (Norheim et al., 2014). In conclusion, ANGPTL4 is highly induced in muscle in response to exercise. However, adipose tissue and the liver may contribute more than muscle to the exercise-induced increase in circulating ANGPTL4. This protein may even be considered as an adipohepatomyokine. This highlights the complexity of secreted factors in the context of different organs and the need of an integrated view to better understand the individual function of certain components of the cellular secretome.

The chemokine monocyte chemoattractant protein 1 (MCP-1) is another example of an adipomyokine that is regulated by exercise in muscle. Among other functions, it plays an important role in the recruitment of monocytes and T lymphocytes into tissues (Qin et al., 1996) and plays a key role in inflammation. Interestingly, acute resistance exercise strongly increased MCP-1 mRNA expression in skeletal muscle of young male volunteers 2 h postexercise (Vella et al., 2012). These results are supported by another study that observed a significant increase in MCP-1 protein level in skeletal muscle after acute resistance training in young and elderly healthy subjects (Della Gatta et al., 2014). Furthermore, cycling at 70% VO_{2max} for 40 min enhanced MCP-1 mRNA expression in skeletal muscle of lean, obese, and type 2 diabetic subjects (Tantiwong et al., 2010). An additional study found increased MCP-1 plasma levels after acute treadmill running at different intensities with significantly higher plasma levels of MCP-1 after the high-intensity compared with the moderate-intensity trial (Peake et al., 2005). These results suggest that the exercise-induced production of MCP-1 is more influenced by the intensity of exercise than by exercise-induced muscle damage, but it is not clear what role MCP-1 plays under exercising conditions. With regard to chronic resistance training, no effect on MCP-1 protein expression in the muscle of young and elderly healthy subjects was observed after 12 weeks (Della Gatta et al., 2014). Moreover, plasma MCP-1 levels were not changed after 12 weeks of low-intensity resistance training in a study with sedentary, lean, elderly women (Ogawa et al., 2010). It is worth noting that other studies observed a decrease in MCP-1 plasma levels after chronic strength training. This could be related to the observed reduction of body fat mass under these conditions, highlighting again the overlap between tissues and their secreted factors.

The proinflammatory molecule TNFα is an early mediator of local inflammatory responses as well as initiator of the systemic acute phase response. It is produced by adipose tissue, and circulating TNFα levels are positively correlated with obesity and the hypertrophy of adipose tissue. This aspect has been discussed in detail in Chapter 2. However, TNFα mRNA is also detectable in skeletal muscle, but no difference in its expression was found between overweight and lean subjects (Plomgaard et al., 2007) or type 2 diabetics and BMI-matched controls. Importantly, it was demonstrated that TNFα is not released by skeletal muscle after acute exercise in either healthy subjects or patients with type 2 diabetes (Febbraio et al., 2003). Furthermore, chronic resistance or endurance training significantly reduced TNFα mRNA and protein in skeletal muscle but had no effect on circulating TNFα concentrations. More importantly, a large body of evidence shows an inverse relationship between plasma TNFα levels and the amount of physical activity even in healthy lean subjects as reviewed by Golbidi and Laher (2014). In contrast to the observation that regular moderate exercise reduces proinflammatory cytokines such as TNFα, high-intensity training is known to induce a temporary depression of various aspects of immune function and an increase in systemic inflammation for a certain postexercise period (~3–24 h). TNFα levels increase only during very intensive exercise (such as marathon running) in response to muscle damage (Paulsen et al., 2012). It is well established that muscle repair and regeneration after acute muscle injury involves a tissue-remodeling, growth-promoting local inflammation. Thus, proinflammatory adipomyokines may play an important role for the plasticity of muscle tissue under different physiological conditions.

Nicotinamide phosphoribosyl transferase (NAMPT) is a ubiquitously expressed nicotinamide dinucleotide (NAD) biosynthetic enzyme that occurs either in an intracellular (iNAMPT) or extracellular form (eNAMPT/visfatin) (Costford et al., 2010). In mammals, NAMPT is responsible for the first and rate-limiting step in the conversion of nicotinamide to NAD^+ in the NAD^+ salvage pathway. Costford et al. (2010) reported a twofold higher NAMPT expression in skeletal muscle of athletes as compared with sedentary obese, nonobese, and type 2 diabetic subjects. Moreover, 3 weeks of endurance training enhanced NAMPT mRNA expression and protein content more than twofold in skeletal muscle of nonobese sedentary individuals. A study by Brandauer et al. (2013) using one-legged endurance exercise training for 3 weeks confirmed these results and reported a specific upregulation of NAMPT in the exercising leg. Interestingly, acute exercise using 3 h of cycling at 60% VO_{2max} had no effect on skeletal muscle NAMPT expression, and circulating eNAMPT/visfatin was found to be unchanged. However, an acute bout of high-intensity exercise was shown to increase plasma eNAMPT/visfatin immediately after the challenge (Ghanbari-Niaki et al., 2010), pointing towards the notion that acute release of eNAMPT/visfatin could be intensity-dependent. It was also reported that skeletal muscle NAMPT protein was negatively correlated with body fat, whereas in obesity and type 2 diabetes circulating eNAMPT/visfatin was found to be elevated compared with controls (Costford et al., 2010). In addition, several chronic exercise intervention studies conducted with obese volunteers reported reductions in plasma eNAMPT/visfatin levels. These observations suggest a potential impact of the adipose tissue on circulating levels of eNAMPT/visfatin and an independent regulation of intracellular NAMPT and circulating eNAMPT/visfatin, which awaits further investigations, specifically with regard to the effect of physical activity.

Although adiponectin is a classical adipokine, it is also expressed in skeletal muscle (Delaigle et al., 2004) and this represents an adipomyokine. In obesity and insulin resistance, plasma adiponectin levels are lower, whereas skeletal muscle expression of adiponectin receptors AdipoR1 and AdipoR2 is increased (Bluher et al., 2006). These authors found that chronic endurance training increased both plasma adiponectin levels as well as expression of AdipoR1 and AdipoR2 in skeletal muscle of overweight/obese subjects with impaired glucose metabolism and in lean healthy controls. However, in severely obese subjects a combination of endurance exercise and diet increased plasma adiponectin levels but had no effect on AdipoR1 and AdipoR2 mRNA expression in skeletal muscle (Bruun et al., 2006). So far, only few studies have investigated the regulation of adiponectin expression in skeletal muscle upon various types of exercise in different groups of patients, and a systematic review on this topic was published recently (Yu et al., 2017).

Also leptin is a classical adipokine but has been found to be expressed and released by skeletal muscle (Wolsk et al., 2012). These authors reported that leptin is released by human skeletal muscle in vivo and found a release of ~ 0.8 ng per min per 100 g tissue from adipose tissue and ~ 0.5 ng per min per 100 g tissue from skeletal muscle. These data suggest that the contribution of skeletal muscle to whole-body leptin production could be substantial in lean humans because of the greater muscle mass compared with fat mass.

EXERCISE AND THE PRODUCTION OF ADIPOKINES BY ADIPOSE TISSUE

Although numerous studies have investigated the effect of different types of exercise training on circulating levels of adipose tissue—derived factors (Sakurai et al. 2013, 2017; Garcia-Hermoso et al., 2017; Duzova et al., 2016; van Gemert et al., 2016; Lee and Kang, 2015; Bluher et al., 2014; Phillips and Cobbold, 2014; Santa Mina et al., 2013), only few studies have addressed the adipose tissue—specific expression of these factors. Adiponectin is one of the best characterized classical adipokines (see Chapter 2) and has been shown to increase fatty acid oxidation and glucose uptake in skeletal muscle and inhibit gluconeogenesis in the liver. In macrophages, adiponectin inhibits expression and secretion of TNFα while increasing the production of antiinflammatory cytokines such as IL-10. Adiponectin is known to exert cardioprotective effects and is inversely related to BMI (Han et al., 2007). Studies on the regulation of adiponectin by exercise have shown divergent results, although most studies indicate no effect of exercise. Two reports have investigated adiponectin mRNA expression in adipose tissue after acute exercise and found either increased (Christiansen et al., 2013) or decreased (Hojbjerre et al., 2007) expression after cycling in overweight, obese, and lean subjects with no difference between the groups. Also, plasma adiponectin levels remained unchanged after acute exercise with moderate intensity in healthy (Ferguson et al., 2004) or overweight/obese patients (Bouassida et al., 2010). Considering chronic exercise training, most studies using endurance-type training with moderate intensity reported no change in plasma levels of adiponectin, whereas mRNA expression in adipose tissue was found to be either increased or unmodified (Christiansen et al., 2010; Hulver et al., 2002; Moghadasi et al., 2013; Polak et al., 2006). However, two other studies that also used chronic endurance exercise protocols reported increased plasma adiponectin levels in obese young women and in overweight patients with impaired glucose metabolism after the intervention period (Kondo et al., 2006). Thus, it appears that endurance exercise training is not a major regulator of adipose tissue adiponectin expression and release, although more work will be required to provide a definitive answer to this topic. In studies using resistance exercise for chronic intervention, the expression of adiponectin mRNA in adipose tissue as well as circulating levels was found to be unchanged in obese men and women (Klimcakova et al., 2006; Phillips et al., 2012). However, it is important to note that in overweight elderly men resistance training at high intensity for 6 months is able to augment plasma adiponectin levels, whereas a moderate-intensity training had no effect (Fatouros et al., 2005). Similar results were reported for acute resistance exercise in overweight elderly men (Fatouros et al., 2009). These data suggest that the intensity of the exercise program may be an important determinant in the regulation of adiponectin concentration in adipose tissue and in plasma.

Leptin is a well-established classical adipokine predominantly secreted from adipocytes into the circulation to regulate energy homeostasis and a number of other physiological processes (Allison and Myers, 2014; Irving and Harvey, 2014; Londraville et al., 2014; McGregor and Harvey, 2017; Park and Ahima, 2015; Roumaud and Martin, 2015). In Chapter 2, we have already described the role of leptin in the crosstalk from adipose tissue to other organs. Interestingly, leptin is also regulated by physical activity, albeit with some divergent results. After acute endurance exercise, leptin mRNA in adipose tissue was found to be unchanged (Keller et al., 2005) or decreased in lean and overweight subjects (Hojbjerre et al., 2007). Moreover, plasma leptin levels analyzed immediately after the exercise challenge did not show any differences compared with preexercise levels (Perusse et al., 1997; Varady et al., 2010), whereas other groups found a delayed decrease of circulating leptin levels in healthy active men 24 and 48 h postexercise (Essig et al., 2000; Olive and Miller, 2001; Yang et al., 2014). After chronic intervention for ≥ 12 weeks, leptin mRNA expression in adipose tissue remained unaltered after an endurance training program performed by obese subjects (Polak et al., 2006). In most of the studies, not only leptin plasma concentrations were found to be decreased (O'Leary et al., 2006), but also body weight and fat mass of the participants were reduced after the intervention periods. Nevertheless, Polak et al. (2006) reported declined leptin plasma levels, which remained significant after adjusting for BMI and fat mass, suggesting independent effects of the training besides reduction of body weight. Moreover, studies using resistance training for chronic exercise intervention reported decreased leptin plasma levels without changes in body weight and fat mass and no effect on adipose tissue mRNA expression of leptin (Phillips et al., 2012). So far, the available data suggest that plasma leptin levels are decreased mainly because of chronic intervention, whereas adipose tissue mRNA expression seems to be not affected by exercise.

TNFα is a major inflammatory cytokine that is highly expressed in adipose tissue in obese conditions and plays a role in the pathogenesis of insulin resistance, as described in Chapter 2. Some recent data highlight the impact of adipose tissue TNFα (Lorente-Cebrian et al., 2014; Nair et al., 2013; Palacios-Ortega et al., 2015). After acute exercise, TNFα mRNA expression in adipose tissue was found to be unchanged immediately after the bout of exercise in overweight, obese, and lean subjects. However, a higher TNFα mRNA expression level was observed in the postexercise period 2.5 h after completing the exercise bout (Christiansen et al., 2013). Concerning the circulating TNFα levels, either no effect of acute exercise was found (Harris et al., 2008) or increased plasma TNFα concentrations were reported. Several studies on chronic exercise using either endurance or resistance training found no effect on adipose tissue mRNA expression of TNFα (Klimcakova et al., 2006). The TNFα levels in the circulation were mostly reported to be reduced by chronic exercise (Phillips et al., 2012). Overall, chronic exercise seems to be able to reduce TNFα levels without an effect on adipose tissue expression.

As discussed earlier, IL-6 is the prototype of an adipomyokine. As described in detail in Chapter 3, it is well established that the increase of circulating IL-6 upon exercise is mainly due to its release from skeletal muscle. However, subjects with insulin resistance, obesity, and type 2 diabetes display chronically elevated serum levels of IL-6. An important source of circulating IL-6 in obesity is the expanding visceral adipose tissue mass (Hunter and Jones, 2015; Mauer et al., 2015). Expression of IL-6 by macrophages within the adipose tissue is dependent on the activation of the NFκB signaling pathway, whereas intramuscular IL-6 expression is regulated by different signaling cascades. Hojbjerre et al. (2007) have analyzed microdialysates from abdominal adipose tissue in overweight and lean males to clarify a potential contribution of adipose tissue to increased IL-6 levels after acute endurance exercise. Interestingly, they observed increased IL-6 release from adipose tissue in the postexercise phase but not during exercise in both groups. Moreover, IL-6 mRNA expression in adipose tissue was increased during and post exercise. Comparable data were published by Christiansen et al. (2013). After chronic endurance training with moderate weight loss, no reduction of IL-6 mRNA expression in adipose tissue of obese subjects was found and basal IL-6 concentrations remained unaffected (Polak et al., 2006). In the case of chronic resistance exercise, no reduction of adipose tissue IL-6 mRNA expression or basal circulating IL-6 in obese subjects was observed (Phillips et al., 2012). Thus, compared with the very prominent effect of exercise on muscle IL-6, the release of IL-6 from adipose tissue and its contribution to the circulating level of this protein appear to be mostly regulated by adipose tissue mass and most likely the level of adipose tissue inflammation and infiltration by macrophages.

In addition to the previously discussed adipokines/adipomyokines, only few studies have investigated other adipose tissue–derived factors in response to exercise. Resistin is an inflammatory biomarker and potential mediator of obesity-associated diseases. It is positively correlated to percent fat mass and waist circumference, and evidence suggests that this adipokine causes endothelial dysfunction by promoting oxidative stress and downregulating the production of nitric oxide (Codoner-Franch and Alonso-Iglesias, 2015). Acute endurance training had no effect on circulating resistin levels up to 48 h postexercise in overweight men (Jamurtas et al., 2006). In line with these findings, in another study with lean and overweight volunteers, resistin mRNA expression in adipose tissue was not affected (Hojbjerre et al., 2007). A lifestyle intervention combining chronic endurance training and hypocaloric diet with obese patients for 3 months also had no effect on plasma resistin concentrations (de Luis et al., 2006). Interestingly, a study using acute resistance exercise observed that the exercise effect on plasma resistin levels was depending on the training status of the subjects. Although a reduction was observed in participants who performed regular weight training (≥ 1 h, three times/week, for 6 months before the study), no effect was found in sedentary males or in active runners (running ≥ 15 miles/week for 6 months before the study) (Varady et al., 2010).

With regard to MCP-1, an important proinflammatory adipomyokine, studies have found no effect of chronic endurance training on adipose tissue mRNA expression despite reductions of circulating MCP-1 levels (Christiansen et al., 2010). Also, chronic resistance training in obese women did not modify MCP-1 mRNA expression in adipose tissue. Observed changes in circulating MCP-1 may therefore be more related to a loss of adipose tissue mass and reduced adipose tissue inflammation.

NAMPT/visfatin is an interesting adipokine with multiple biological functions discussed in detail in Chapter 2. So far, only one study has investigated adipose tissue NAMPT expression after exercise and reported an increase after an acute bout of endurance training, which was not accompanied by elevated plasma eNAMPT/visfatin concentrations. Reports on the effect of chronic exercise on circulating eNAMPT/visfatin levels are not consistent. In obese patients, higher eNAMPT/visfatin levels compared with controls were observed, which were reduced by 12 weeks of aerobic or aerobic plus resistance exercise accompanied by decreased body weight (Lee et al., 2010). Based on these data it is most likely that exercise-induced reduction of plasma eNAMPT/visfatin is the result of weight loss and body composition changes.

Few data are available for ANGPTL4 in human adipose tissue after exercise. In a study by Norheim et al. (2014), it was reported that 30 min after an acute bout of endurance exercise ANGPTL4 mRNA expression was significantly higher in overweight dysglycemic subjects compared with that in lean controls at the beginning of the intervention study. In this study it was observed that ANGPTL4 mRNA expression was much higher in adipose tissue than muscle. After 12 weeks of combined endurance and resistance training, adipose tissue mRNA expression of ANGPTL4 as well as basal plasma levels was unchanged in both groups. In contrast, another chronic endurance training study in obese but otherwise healthy participants found significantly reduced mRNA expression in adipose tissue but increased circulating ANGPTL4 levels (Cullberg et al., 2013). Clearly, more studies are required to understand the regulation of ANGPTL4 in response to various types of exercise in different types of subjects as well as the contribution of the different tissues to circulating ANGPTL4. In fact, in a very recent study it was shown that ANGPTL4 is an exercise-induced hepatokine in humans (Ingerslev et al., 2017). These authors investigated the origin of exercise-induced ANGPTL4 in humans by measuring the arterial-to-venous difference over the leg and the hepatosplanchnic bed during an acute bout of exercise. Furthermore, the impact of the glucagon-to-insulin ratio on plasma ANGPTL4 was studied. The regulation of ANGPTL4 was investigated in both hepatic and muscle cells. The data of this study suggest that the glucagon-to-insulin ratio is an important regulator of plasma ANGPTL4, as elevated glucagon in the absence of elevated insulin increased plasma ANGPTL4 in resting subjects. Moreover, activation of the cAMP/PKA signaling cascade led to an increase in ANGPTL4 mRNA levels in hepatic cells, which was prevented by inhibition of PKA. In humans, muscle ANGPTL4 mRNA increased during fasting, with only a marginal further induction by exercise. The authors suggest that exercise-induced

ANGPTL4 is secreted from the liver and driven by a glucagon–cAMP–PKA pathway in humans. These findings link the liver, insulin/glucagon, and lipid metabolism together, which could implicate a role of ANGPTL4 in metabolic diseases.

EXERCISE-INDUCED ADIPOMYOKINES AND THEIR ROLE IN BONE METABOLISM

In Chapter 3, we have already briefly addressed the "muscle–bone" connection, which refers to the close functional and developmental relationship between muscle and bone. The so-called bone–muscle unit concept is based on the lifelong association between lean muscle mass and the bone mineral content and reflects the tight functional association of these two tissues, which is mediating locomotion (Ferretti et al., 1998; Rauch et al., 2004). Many studies have demonstrated that different physio- and pathophysiological stimuli exert a concomitant effect on both bone and muscle, supporting the notion that the musculoskeletal system is orchestrated by a multidirectional crosstalk system. In addition to myokines, a number of osteokines (IGF-1, OCN, PGE2, and others) produced by both osteoblasts and osteocytes were recently described and are thought to affect muscle development (Dallas et al., 2013). However, most likely a much more complex organ crosstalk scenario is taking place at this level and additional tissues such as cartilage, adipose tissue, and tendons need to be integrated to understand the muscle–bone unit. Fig. 4 in Chapter 3, which is based on a recent review by Tagliaferri et al. (2015), provides a schematic view on this multiorgan crosstalk setting. Bone formation is regulated by both mechanical loading and the paracrine effect of secreted myokines (Hamrick et al., 2010). We have already briefly discussed bone-related functions of key myokines such as IL-6 (Hamrick, 2011), IGF-1 (Banu et al., 2003), and myostatin (Elkasrawy et al., 2012). Here we zoom in on the exercise-mediated regulation of adipomyokines and the relation to bone metabolism. This scenario was addressed in an excellent recent review by Lombardi et al. (2016). A number of molecules secreted by muscle and adipose tissues in response to exercise are involved in the fine regulation of bone metabolism in response to the energy availability. Furthermore, bone regulates energy metabolism by communicating its energetic needs, thanks to osteocalcin that acts on pancreatic β-cells and adipocytes. Many of these molecules are adipomyokines, and the overlap between these molecules at the level of bone is a paradigm for the complexity of organ crosstalk. The beneficial effects of exercise on bone metabolism depend on the intermittent exposure to myokines (i.e., IL-6, LIF, IGF-I), which, instead, act as inflammatory/proresorptive mediators when chronically elevated; on the other hand, the reduction in the circulating levels of adipokines (i.e., leptin, visfatin, adiponectin, resistin) sustains these effects as well as improves the whole-body metabolic status.

Physical exercise and bone metabolism are tightly coupled depending on different parameters such as the type of exercise, intensity, duration, and, most importantly, load degree (Banfi et al., 2010). Although the mechanical load applied to the skeleton is a major determinant of bone mass, the endocrine crosstalk in response to exercise involving several tissues has raised considerable interest and may play an important role in the pathophysiology of bone disorders (DiVasta and Gordon, 2013). Despite numerous studies, the optimal intensity, duration, and frequency of physical exercise to optimize bone formation has remained controversial. Data suggest that high strain rates and high peak forces are most powerful mediators of bone formation and are also able to limit bone loss in the elderly (Mosti et al., 2014).

It is important to note that the muscle—bone crosstalk already starts in fetal life. This was evidenced in studies of knockout mice, which congenitally lack skeletal muscle leading to death after birth. These animals showed low mineralization and morphological bone alterations (Gomez et al., 2007). The fine-tuning of the myostatin system is also an important determinant of bone mass. Exercise is known to regulate proteins such as FST, Fstl1, and Fstl3, inhibitors of the myostatin system. The gold standard of a myokine, IL-6, has already been discussed in Chapter 3. In fact, it is widely accepted that the basal level of IL-6 is reduced in response to physical exercise and this correlates with the beneficial effects of exercise on bone (Lombardi et al., 2016). High levels of IL-6 have a negative effect on bone metabolism, as shown in IL-6 transgenic mice that have decreased osteoblasts and increased osteoclasts and develop osteoporosis (De Benedetti et al., 2001).

A number of beneficial effects of physical exercise are dependent on the anabolic effects of IGF-1 on muscle and bone. Myocytes are known to express IGF-1, and data show that this can be localized to the muscle—bone interface, showing also a high level of IGF-1 receptors (Hamrick et al., 2010). Indeed, GH, IGF-1, and IGF-2 act as key regulators of bone homeostasis (Giustina et al., 2008). At present, the contribution of systemic versus locally produced IGF-1 is not clear. The scenario is complicated by the regulation of freely available IGF-1 by different IGFBPs (IGF binding proteins), also expressed by osteoblasts. Furthermore, exercise increases both IGF-1 and IGFBP-3 (Ehrnborg et al., 2003).

As pointed out before in this chapter, physical exercise also affects the adipokine expression profile of adipose tissue. These adipomyocytokines also exert a prominent effect on bone metabolism, highlighting the complexity of the physiological responses toward exercise. Table 4.1 summarizes some myokines and adipokines with proven effects on bone metabolism (for details, see Lombardi et al., 2016). Leptin is well known to be detrimental for bone metabolism, partially owing to its proinflammatory action profile. Because exercise reduces the level of circulating leptin, this also has to be considered as a beneficial effect on bone metabolism. Interestingly, adiponectin levels appear to be negatively related to bone mineral density (Biver et al., 2011). However, conflicting data exist regarding adiponectin as a biomarker of osteoporosis (Lubkowska et al., 2014), and data suggest that adiponectin acts both directly and indirectly on osteoblasts and osteoclasts (Shinoda et al., 2006). As described earlier, exercise decreases circulating visfatin; however, controversial data exist regarding a possible association between physical activity, visfatin, and bone status. Finally, resistin has gained interest as a player in bone metabolism. Data show that osteoclasts and osteoblasts express resistin and that it may stimulate bone remodeling by augmenting osteoclastogenesis and osteoblast

Table 4.1 Effects of Exercise-Induced Myokines and Adipokines Release on Bone Tissue

Molecules	Tissue of Origin	Effects of Exercise	Effects on Bone
Myokines			
IL-6	Liver	Exercise ↓ chronic proinflammatory levels	↓ Inflammation
	Immune cells		↑ Bone status
	Skeletal muscle	Strong exercise ↑ ↑	↑ Bone turnover markers
IL-7	Skeletal muscle	Strength training ↑	Similar to muscle-derived IL-6
	Immune system	Training ↑	Indirect activation of osteoclast differentiation (through immune cells)
IL-8	Monocytes and macrophages	Strenuous endurance activities ↑	↑ Bone resorption
LIF	Working muscles	Mild exercise ↑	↑ Bone turnover
IGF-I	Liver	Strong exercise ↑	↑ NTx, TRAP-5b, and OC
	Extrahepatic tissue		↑ Bone anabolism
Irisin	Skeletal muscle	Exercise ↑	"Browning" of white adipose tissue with consequent enhancement of thermogenesis
BDNF	Neuronal tissues	Exercise ↑	Expressed by osteoblasts during fracture healing
	Myocytes		
Adipokines			
Leptin	Adipose tissue	Inactivity ↑	↑ Proinflammatory cytokines
			↑ Osteoclast activity
		Physical activity ↓	↑ Total OC and GluOC
			↑ Insulin sensitivity
			↓ Fat mass
Adiponectin	Adipose tissue	Regular exercise ↑	Autocrine/paracrine action enhanced osteoblast stimulation
		Single bout has no effect	Systemic action increased osteoclastogenesis

Table 4.1 Effects of Exercise-Induced Myokines and Adipokines Release on Bone Tissue—cont'd

Molecules	Tissue of Origin	Effects of Exercise	Effects on Bone
Visfatin	Skeletal muscle	Exercise ↑ intracellular form	↑ NAD$^+$
	Adipose tissue	Exercise ↓ extracellular form	↑ Insulin sensitivity
Resistin	Adipose tissue	Aerobic activity and endurance ↓	↓ Bone remodeling

Summary of myokine and adipokine changes during physical activity and their effects on bone metabolism.
BDNF, brain-derived neurotrophic factor; *GluOC,* undercarboxylated osteocalcin; *IGF-I,* insulin-like growth factor I; *LIF,* leukemia inhibitory factor; *NTx-I,* N-terminal cross-linked telopeptide of type I collagen; *OC,* osteocalcin; *TRAP-5b,* tartrate-resistant acid phosphatase 5b.
Reproduced with permission from Lombardi, G., Sanchis-Gomar, F., Perego, S., Sansoni, V., Banfi, G., 2016. Implications of exercise-induced adipo-myokines in bone metabolism. Endocrine 54, 284–305.

proliferation (Thommesen et al., 2006). At present, no data are available regarding the exercise-dependent changes of resistin and a possible role in bone metabolism.

CONCLUSION AND FUTURE DIRECTIONS

There are hundreds of secretory proteins released from skeletal muscle as well as adipose tissue with an array of biological effects on most organs in the body. It is a striking observation that many of the myokines and adipokines are expressed and secreted from many different tissues. It seems as if there are very few of these adipomyokines that are exclusively expressed in one organ. However, considering that both adipose tissue and skeletal muscle tissues are closely associated to strong phenotypic traits, it is likely that the pattern of signal molecules released from these two tissues has marked physiological effects of essential importance for health as well as well-being. The original expectation that we would be able to describe discrete signatures of secretory proteins from muscle and adipose will have to be replaced by hard work and acceptance of a much more complicated interplay than many had expected.

The result of a single bout of exercise is predominantly the secretion of myokines and adipomyokines by the working skeletal muscle, which exert a variety of auto-crine and endocrine effects (Fig. 4.2). Acute induction of myokines such as myosta-tin, IL-7, decorin, and LIF is involved in the regulation of muscle growth and may play a role in exercise-related restructuring of skeletal muscle. Muscle-derived IL-6 has metabolic effects, by affecting insulin-stimulated glucose disposal and fatty acid oxidation. Furthermore, acute high levels of circulating IL-6 provide an antiinflam-matory environment after exercise by induction of IL-1ra and IL-10 and inhibition of TNFα production. One the other hand, regular exercise training is associated with

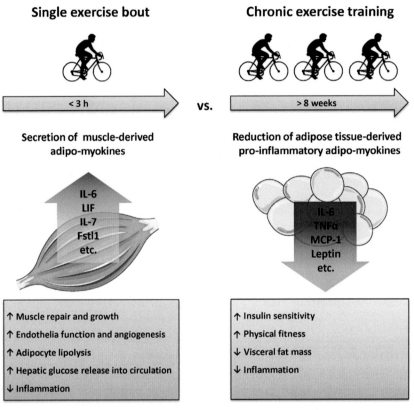

FIGURE 4.2

Differential effects of acute and chronic exercise. After acute exercise, a high number of myokines are secreted by skeletal muscle exerting a variety of endocrine effects. Acute induction of myokines such as myostatin, IL-7, decorin, and LIF is involved in the regulation of muscle hypertrophy and may play a role in exercise-related restructuring of skeletal muscle. The high level of circulating IL-6 after exercise induces an antiinflammatory environment by inducing the production of IL-1ra and IL-10 and also inhibits TNFα production. Furthermore, IL-6 has metabolic effects, by affecting insulin-stimulated glucose disposal and fatty acid oxidation. The myokine Fstl1 has protective effects on ischemia—reperfusion injury in muscle and heart tissue. On the other hand, regular exercise training induces a reduction of adipose tissue—derived proinflammatory cytokines such as IL-6, TNFα, and MCP-1, which are associated with low-grade systemic inflammation and a reduction of whole-body insulin sensitivity. Exercise training has been reported to reduce central adiposity independent of overall weight changes, and this may represent an additional mechanism of the antiinflammatory action of chronic exercise training. *Fstl1*, follistatin-like 1; *LIF*, leukemia inhibitory factor; *MCP-1*, monocyte chemoattractant protein 1.

Reproduced with permission from Goergens, S., et al., 2015. Exercise and regulation of adipokine and myokine production. In: Claude Bouchard, editor, Progress in Molecular Biology and Translational Science. vol. 135. Academic Press, Burlington, pp. 313—336.

reduced levels of adipose tissue—derived proinflammatory cytokines, which are linked with low-grade systemic inflammation and low whole-body insulin sensitivity (Fig. 4.2). Chronic exercise reduces visceral fat mass and has also been reported to reduce central adiposity independent of overall weight changes. The antiinflammatory effects of regular exercise training may be mediated by both the reduction of body fat mass and the induction of an antiinflammatory environment with each single bout of exercise. In summary, regular exercise reduces the risk of chronic metabolic diseases, and various mechanisms may contribute to this beneficial effect, including decreased production of adipose tissue—derived proinflammatory cytokines and increased production of antiinflammatory myokines from contracting muscle.

REFERENCES

Allison, M.B., Myers Jr., M.G., 2014. 20 years of leptin: connecting leptin signaling to biological function. J. Endocrinol. 223, T25—T35.

Amiri Arimi, S., Mohseni Bandpei, M.A., Javanshir, K., Rezasoltani, A., Biglarian, A., 2017. The effect of different exercise programs on size and function of deep cervical flexor muscles in patients with chronic nonspecific neck pain: a systematic review of randomized controlled trials. Am. J. Phys. Med. Rehabil. 96, 582—588.

Banfi, G., Lombardi, G., Colombini, A., Lippi, G., 2010. Bone metabolism markers in sports medicine. Sports Med. 40, 697—714.

Banu, J., Wang, L., Kalu, D.N., 2003. Effects of increased muscle mass on bone in male mice overexpressing IGF-I in skeletal muscles. Calcif. Tissue Int. 73, 196—201.

Biver, E., Salliot, C., Combescure, C., Gossec, L., Hardouin, P., Legroux-Gerot, I., Cortet, B., 2011. Influence of adipokines and ghrelin on bone mineral density and fracture risk: a systematic review and meta-analysis. J. Clin. Endocrinol. Metab. 96, 2703—2713.

Bluher, M., Bullen Jr., J.W., Lee, J.H., Kralisch, S., Fasshauer, M., Kloting, N., Niebauer, J., Schon, M.R., Williams, C.J., Mantzoros, C.S., 2006. Circulating adiponectin and expression of adiponectin receptors in human skeletal muscle: associations with metabolic parameters and insulin resistance and regulation by physical training. J. Clin. Endocrinol. Metab. 91, 2310—2316.

Bluher, S., Panagiotou, G., Petroff, D., Markert, J., Wagner, A., Klemm, T., Filippaios, A., Keller, A., Mantzoros, C.S., 2014. Effects of a 1-year exercise and lifestyle intervention on irisin, adipokines, and inflammatory markers in obese children. Obesity 22, 1701—1708.

Bomb, R., Heckle, M.R., Sun, Y., Mancarella, S., Guntaka, R.V., Gerling, I.C., Weber, K.T., 2016. Myofibroblast secretome and its auto-/paracrine signaling. Expert Rev. Cardiovasc Ther. 14, 591—598.

Bouassida, A., Lakhdar, N., Benaissa, N., Mejri, S., Zaouali, M., Zbidi, A., Tabka, Z., 2010. Adiponectin responses to acute moderate and heavy exercises in overweight middle aged subjects. J. Sports Med. Phys. Fitness 50, 330—335.

Brandauer, J., Vienberg, S.G., Andersen, M.A., Ringholm, S., Risis, S., Larsen, P.S., Kristensen, J.M., Frosig, C., Leick, L., Fentz, J., Jorgensen, S., Kiens, B., Wojtaszewski, J.F., Richter, E.A., Zierath, J.R., Goodyear, L.J., Pilegaard, H., Treebak, J.T., 2013. AMP-activated protein kinase regulates nicotinamide phosphoribosyl transferase expression in skeletal muscle. J. Physiol. 591, 5207—5220.

Broholm, C., Laye, M.J., Brandt, C., Vadalasetty, R., Pilegaard, H., Pedersen, B.K., Scheele, C., 2011. LIF is a contraction-induced myokine stimulating human myocyte proliferation. J. Appl. Physiol. 111, 251–259.

Bruun, J.M., Helge, J.W., Richelsen, B., Stallknecht, B., 2006. Diet and exercise reduce low-grade inflammation and macrophage infiltration in adipose tissue but not in skeletal muscle in severely obese subjects. Am. J. Physiol. Endocrinol. Metab. 290, E961–E967.

Carson, B.P., 2017. The potential role of contraction-induced myokines in the regulation of metabolic function for the prevention and treatment of type 2 diabetes. Front. Endocrinol. 8, 97.

Catoire, M., Alex, S., Paraskevopulos, N., Mattijssen, F., Evers-van Gogh, I., Schaart, G., Jeppesen, J., Kneppers, A., Mensink, M., Voshol, P.J., Olivecrona, G., Tan, N.S., Hesselink, M.K., Berbee, J.F., Rensen, P.C., Kalkhoven, E., Schrauwen, P., Kersten, S., 2014. Fatty acid-inducible ANGPTL4 governs lipid metabolic response to exercise. Proc. Natl. Acad. Sci. U. S. A. 111, E1043–E1052.

Chen, P.R., Lee, K., 2016. Invited review: inhibitors of myostatin as methods of enhancing muscle growth and development. J. Anim. Sci. 94, 3125–3134.

Christiansen, T., Paulsen, S.K., Bruun, J.M., Pedersen, S.B., Richelsen, B., 2010. Exercise training versus diet-induced weight-loss on metabolic risk factors and inflammatory markers in obese subjects: a 12-week randomized intervention study. Am. J. Physiol. Endocrinol. Metab. 298, E824–E831.

Christiansen, T., Bruun, J.M., Paulsen, S.K., Olholm, J., Overgaard, K., Pedersen, S.B., Richelsen, B., 2013. Acute exercise increases circulating inflammatory markers in overweight and obese compared with lean subjects. Eur. J. Appl. Physiol. 113, 1635–1642.

Codoner-Franch, P., Alonso-Iglesias, E., 2015. Resistin: insulin resistance to malignancy. Clin. Chim. Acta 438, 46–54.

Costford, S.R., Bajpeyi, S., Pasarica, M., Albarado, D.C., Thomas, S.C., Xie, H., Church, T.S., Jubrias, S.A., Conley, K.E., Smith, S.R., 2010. Skeletal muscle NAMPT is induced by exercise in humans. Am. J. Physiol. Endocrinol. Metab. 298, E117–E126.

Cullberg, K.B., Christiansen, T., Paulsen, S.K., Bruun, J.M., Pedersen, S.B., Richelsen, B., 2013. Effect of weight loss and exercise on angiogenic factors in the circulation and in adipose tissue in obese subjects. Obesity 21, 454–460.

Dallas, S.L., Prideaux, M., Bonewald, L.F., 2013. The osteocyte: an endocrine cell... and more. Endocr. Rev. 34, 658–690.

De Benedetti, F., Pignatti, P., Vivarelli, M., Meazza, C., Ciliberto, G., Savino, R., Martini, A., 2001. In vivo neutralization of human IL-6 (hIL-6) achieved by immunization of hIL-6-transgenic mice with a hIL-6 receptor antagonist. J. Immunol. 166, 4334–4340.

de Luis, D.A., Aller, R., Izaola, O., Sagrado, M.G., Conde, R., 2006. Influence of ALA54THR polymorphism of fatty acid binding protein 2 on lifestyle modification response in obese subjects. Ann. Nutr. Metab. 50, 354–360.

Delaigle, A.M., Jonas, J.C., Bauche, I.B., Cornu, O., Brichard, S.M., 2004. Induction of adiponectin in skeletal muscle by inflammatory cytokines: in vivo and in vitro studies. Endocrinology 145, 5589–5597.

Della Gatta, P.A., Garnham, A.P., Peake, J.M., Cameron-Smith, D., 2014. Effect of exercise training on skeletal muscle cytokine expression in the elderly. Brain Behav. Immun. 39, 80–86.

Deng, B., Zhang, F., Wen, J., Ye, S., Wang, L., Yang, Y., Gong, P., Jiang, S., 2017. The function of myostatin in the regulation of fat mass in mammals. Nutr. Metab. 14, 29.

DiVasta, A.D., Gordon, C.M., 2013. Exercise and bone: where do we stand? Metabolism 62, 1714–1717.

Duzova, H., Gullu, E., Cicek, G., Koksal, B.K., Kayhan, B., Gullu, A., Sahin, I., 2016. The effect of exercise induced weight-loss on myokines and adipokines in overweight sedentary females: steps-aerobics vs. jogging-walking exercises. J. Sports Med. Phys. Fitness.

Eckardt, K., Gorgens, S.W., Raschke, S., Eckel, J., 2014. Myokines in insulin resistance and type 2 diabetes. Diabetologia 57, 1087–1099.

Ehrnborg, C., Lange, K.H., Dall, R., Christiansen, J.S., Lundberg, P.A., Baxter, R.C., Boroujerdi, M.A., Bengtsson, B.A., Healey, M.L., Pentecost, C., Longobardi, S., Napoli, R., Rosen, T., Study Group, G.H., 2003. The growth hormone/insulin-like growth factor-I axis hormones and bone markers in elite athletes in response to a maximum exercise test. J. Clin. Endocrinol. Metab. 88, 394–401.

Elkasrawy, M., Immel, D., Wen, X., Liu, X., Liang, L.F., Hamrick, M.W., 2012. Immunolocalization of myostatin (GDF-8) following musculoskeletal injury and the effects of exogenous myostatin on muscle and bone healing. J. Histochem. Cytochem. 60, 22–30.

Essig, D.A., Alderson, N.L., Ferguson, M.A., Bartoli, W.P., Durstine, J.L., 2000. Delayed effects of exercise on the plasma leptin concentration. Metabolism 49, 395–399.

Fan, X., Gaur, U., Sun, L., Yang, D., Yang, M., 2017. The growth differentiation factor 11 (GDF11) and myostatin (MSTN) in tissue specific aging. Mech. Ageing Dev. 164, 108–112.

Fatouros, I.G., Tournis, S., Leontsini, D., Jamurtas, A.Z., Sxina, M., Thomakos, P., Manousaki, M., Douroudos, I., Taxildaris, K., Mitrakou, A., 2005. Leptin and adiponectin responses in overweight inactive elderly following resistance training and detraining are intensity related. J. Clin. Endocrinol. Metab. 90, 5970–5977.

Fatouros, I.G., Chatzinikolaou, A., Tournis, S., Nikolaidis, M.G., Jamurtas, A.Z., Douroudos II, I., Papassotiriou, P.M., Thomakos, K., Taxildaris, G., Mastorakos, Mitrakou, A., 2009. Intensity of resistance exercise determines adipokine and resting energy expenditure responses in overweight elderly individuals. Diabetes Care 32, 2161–2167.

Febbraio, M.A., Pedersen, B.K., 2002. Muscle-derived interleukin-6: mechanisms for activation and possible biological roles. FASEB J. 16, 1335–1347.

Febbraio, M.A., Pedersen, B.K., 2005. Contraction-induced myokine production and release: is skeletal muscle an endocrine organ? Exerc. Sport Sci. Rev. 33, 114–119.

Febbraio, M.A., Steensberg, A., Starkie, R.L., McConell, G.K., Kingwell, B.A., 2003. Skeletal muscle interleukin-6 and tumor necrosis factor-alpha release in healthy subjects and patients with type 2 diabetes at rest and during exercise. Metabolism 52, 939–944.

Ferguson, M.A., White, L.J., McCoy, S., Kim, H.W., Petty, T., Wilsey, J., 2004. Plasma adiponectin response to acute exercise in healthy subjects. Eur. J. Appl. Physiol. 91, 324–329.

Ferretti, J.L., Capozza, R.F., Cointry, G.R., Garcia, S.L., Plotkin, H., Alvarez Filgueira, M.L., Zanchetta, J.R., 1998. Gender-related differences in the relationship between densitometric values of whole-body bone mineral content and lean body mass in humans between 2 and 87 years of age. Bone 22, 683–690.

Fischer, C.P., 2006. Interleukin-6 in acute exercise and training: what is the biological relevance? Exerc. Immunol. Rev. 12, 6–33.

Garcia-Hermoso, A., Ceballos-Ceballos, R.J., Poblete-Aro, C.E., Hackney, A.C., Mota, J., Ramirez-Velez, R., 2017. Exercise, adipokines and pediatric obesity: a meta-analysis of randomized controlled trials. Int. J. Obes. 41, 475–482.

Ghanbari-Niaki, A., Saghebjoo, M., Soltani, R., Kirwan, J.P., 2010. Plasma visfatin is increased after high-intensity exercise. Ann. Nutr. Metab. 57, 3–8.

Giudice, J., Taylor, J.M., 2017. Muscle as a paracrine and endocrine organ. Curr. Opin. Pharmacol. 34, 49–55.

Giustina, A., Mazziotti, G., Canalis, E., 2008. Growth hormone, insulin-like growth factors, and the skeleton. Endocr. Rev. 29, 535–559.

Goekint, M., De Pauw, K., Roelands, B., Njemini, R., Bautmans, I., Mets, T., Meeusen, R., 2010. Strength training does not influence serum brain-derived neurotrophic factor. Eur. J. Appl. Physiol. 110, 285–293.

Golbidi, S., Laher, I., 2014. Exercise induced adipokine changes and the metabolic syndrome. J. Diabetes. Res. 2014, 726861.

Gomez, C., David, V., Peet, N.M., Vico, L., Chenu, C., Malaval, L., Skerry, T.M., 2007. Absence of mechanical loading in utero influences bone mass and architecture but not innervation in Myod-Myf5-deficient mice. J. Anat. 210, 259–271.

Gorgens, S.W., Raschke, S., Holven, K.B., Jensen, J., Eckardt, K., Eckel, J., 2013. Regulation of follistatin-like protein 1 expression and secretion in primary human skeletal muscle cells. Arch. Physiol. Biochem. 119, 75–80.

Griffin, E.W., Mullally, S., Foley, C., Warmington, S.A., OMara, S.M., Kelly, A.M., 2011. Aerobic exercise improves hippocampal function and increases BDNF in the serum of young adult males. Physiol. Behav. 104, 934–941.

Hamrick, M.W., 2011. A role for myokines in muscle-bone interactions. Exerc. Sport Sci. Rev. 39, 43–47.

Hamrick, M.W., McNeil, P.L., Patterson, S.L., 2010. Role of muscle-derived growth factors in bone formation. J. Musculoskelet. Neuronal Interact. 10, 64–70.

Han, S.H., Quon, M.J., Kim, J.A., Koh, K.K., 2007. Adiponectin and cardiovascular disease: response to therapeutic interventions. J. Am. Coll. Cardiol. 49, 531–538.

Harber, M.P., Crane, J.D., Dickinson, J.M., Jemiolo, B., Raue, U., Trappe, T.A., Trappe, S.W., 2009. Protein synthesis and the expression of growth-related genes are altered by running in human vastus lateralis and soleus muscles. Am. J. Physiol. Regul. Integr. Comp. Physiol. 296, R708–R714.

Harris, R.A., Padilla, J., Hanlon, K.P., Rink, L.D., Wallace, J.P., 2008. The flow-mediated dilation response to acute exercise in overweight active and inactive men. Obesity 16, 578–584.

Hennigar, S.R., McClung, J.P., Pasiakos, S.M., 2017. Nutritional interventions and the IL-6 response to exercise. FASEB J. 31, 3719–3728.

Hittel, D.S., Axelson, M., Sarna, N., Shearer, J., Huffman, K.M., Kraus, W.E., 2010. Myostatin decreases with aerobic exercise and associates with insulin resistance. Med. Sci. Sports Exerc. 42, 2023–2029.

Hojbjerre, L., Rosenzweig, M., Dela, F., Bruun, J.M., Stallknecht, B., 2007. Acute exercise increases adipose tissue interstitial adiponectin concentration in healthy overweight and lean subjects. Eur. J. Endocrinol. 157, 613–623.

Hulver, M.W., Zheng, D., Tanner, C.J., Houmard, J.A., Kraus, W.E., Slentz, C.A., Sinha, M.K., Pories, W.J., MacDonald, K.G., Dohm, G.L., 2002. Adiponectin is not altered with exercise training despite enhanced insulin action. Am. J. Physiol. Endocrinol. Metab. 283, E861–E865.

Hunter, C.A., Jones, S.A., 2015. IL-6 as a keystone cytokine in health and disease. Nat. Immunol. 16, 448–457.

Ingerslev, B., Hansen, J.S., Hoffmann, C., Clemmesen, J.O., Secher, N.H., Scheler, M., Hrabe de Angelis, M., Haring, H.U., Pedersen, B.K., Weigert, C., Plomgaard, P., 2017. Angiopoietin-like protein 4 is an exercise-induced hepatokine in humans, regulated by glucagon and cAMP. Mol. Metab. 6, 1286−1295.

Irving, A.J., Harvey, J., 2014. Leptin regulation of hippocampal synaptic function in health and disease. Philos. Trans. R. Soc. Lond. B Biol. Sci. 369, 20130155.

Jamurtas, A.Z., Theocharis, V., Koukoulis, G., Stakias, N., Fatouros, I.G., Kouretas, D., Koutedakis, Y., 2006. The effects of acute exercise on serum adiponectin and resistin levels and their relation to insulin sensitivity in overweight males. Eur. J. Appl. Physiol. 97, 122−126.

Jordan, S.C., Choi, J., Kim, I., Wu, G., Toyoda, M., Shin, B., Vo, A., 2017. Interleukin-6, a cytokine critical to mediation of inflammation, autoimmunity and allograft rejection: therapeutic implications of IL-6 receptor blockade. Transplantation 101, 32−44.

Keller, P., Keller, C., Steensberg, A., Robinson, L.E., Pedersen, B.K., 2005. Leptin gene expression and systemic levels in healthy men: effect of exercise, carbohydrate, interleukin-6, and epinephrine. J. Appl. Physiol. 98, 1805−1812.

Kersten, S., Lichtenstein, L., Steenbergen, E., Mudde, K., Hendriks, H.F., Hesselink, M.K., Schrauwen, P., Muller, M., 2009. Caloric restriction and exercise increase plasma ANGPTL4 levels in humans via elevated free fatty acids. Arterioscler. Thromb. Vasc. Biol. 29, 969−974.

Kim, J.S., Cross, J.M., Bamman, M.M., 2005. Impact of resistance loading on myostatin expression and cell cycle regulation in young and older men and women. Am. J. Physiol. Endocrinol. Metab. 288, E1110−E1119.

Klimcakova, E., Polak, J., Moro, C., Hejnova, J., Majercik, M., Viguerie, N., Berlan, M., Langin, D., Stich, V., 2006. Dynamic strength training improves insulin sensitivity without altering plasma levels and gene expression of adipokines in subcutaneous adipose tissue in obese men. J. Clin. Endocrinol. Metab. 91, 5107−5112.

Kohut, M.L., McCann, D.A., Russell, D.W., Konopka, D.N., Cunnick, J.E., Franke, W.D., Castillo, M.C., Reighard, A.E., Vanderah, E., 2006. Aerobic exercise, but not flexibility/resistance exercise, reduces serum IL-18, CRP, and IL-6 independent of beta-blockers, BMI, and psychosocial factors in older adults. Brain Behav. Immun. 20, 201−209.

Kondo, T., Kobayashi, I., Murakami, M., 2006. Effect of exercise on circulating adipokine levels in obese young women. Endocr. J. 53, 189−195.

Lee, S.S., Kang, S., 2015. Effects of regular exercise on obesity and type 2 diabete mellitus in Korean children: improvements glycemic control and serum adipokines level. J. Phys. Ther. Sci. 27, 1903−1907.

Lee, S., Kuk, J.L., Davidson, L.E., Hudson, R., Kilpatrick, K., Graham, T.E., Ross, R., 2005. Exercise without weight loss is an effective strategy for obesity reduction in obese individuals with and without type 2 diabetes. J. Appl. Physiol. (1985) 99, 1220−1225.

Lee, K.J., Shin, Y.A., Lee, K.Y., Jun, T.W., Song, W., 2010. Aerobic exercise training-induced decrease in plasma visfatin and insulin resistance in obese female adolescents. Int. J. Sport Nutr. Exerc. Metab. 20, 275−281.

Lombardi, G., Sanchis-Gomar, F., Perego, S., Sansoni, V., Banfi, G., 2016. Implications of exercise-induced adipo-myokines in bone metabolism. Endocrine 54, 284−305.

Londraville, R.L., Macotela, Y., Duff, R.J., Easterling, M.R., Liu, Q., Crespi, E.J., 2014. Comparative endocrinology of leptin: assessing function in a phylogenetic context. Gen. Comp. Endocrinol. 203, 146−157.

Lorente-Cebrian, S., Mejhert, N., Kulyte, A., Laurencikiene, J., Astrom, G., Heden, P., Ryden, M., Arner, P., 2014. MicroRNAs regulate human adipocyte lipolysis: effects of miR-145 are linked to TNF-alpha. PLoS One 9, e86800.

Lubkowska, A., Dobek, A., Mieszkowski, J., Garczynski, W., Chlubek, D., 2014. Adiponectin as a biomarker of osteoporosis in postmenopausal women: controversies. Dis. Markers 2014, 975178.

Matthews, V.B., Astrom, M.B., Chan, M.H., Bruce, C.R., Krabbe, K.S., Prelovsek, O., Akerstrom, T., Yfanti, C., Broholm, C., Mortensen, O.H., Penkowa, M., Hojman, P., Zankari, A., Watt, M.J., Bruunsgaard, H., Pedersen, B.K., Febbraio, M.A., 2009. Brain-derived neurotrophic factor is produced by skeletal muscle cells in response to contraction and enhances fat oxidation via activation of AMP-activated protein kinase. Diabetologia 52, 1409–1418.

Mattson, M.P., 2012. Energy intake and exercise as determinants of brain health and vulnerability to injury and disease. Cell Metab. 16, 706–722.

Mauer, J., Denson, J.L., Bruning, J.C., 2015. Versatile functions for IL-6 in metabolism and cancer. Trends Immunol. 36, 92–101.

McGregor, G., Harvey, J., 2017. Food for thought: leptin regulation of hippocampal function and its role in Alzheimers disease. Neuropharmacology.

Moghadasi, M., Mohebbi, H., Rahmani-Nia, F., Hassan-Nia, S., Noroozi, H., 2013. Effects of short-term lifestyle activity modification on adiponectin mRNA expression and plasma concentrations. Eur. J. Sport Sci. 13, 378–385.

Mosti, M.P., Carlsen, T., Aas, E., Hoff, J., Stunes, A.K., Syversen, U., 2014. Maximal strength training improves bone mineral density and neuromuscular performance in young adult women. J. Strength Cond. Res. 28, 2935–2945.

Myers, A.M., Beam, N.W., Fakhoury, J.D., 2017. Resistance training for children and adolescents. Transl. Pediatr. 6, 137–143.

Nair, S., Nguyen, H., Salama, S., Al-Hendy, A., 2013. Obesity and the endometrium: adipocyte-secreted proinflammatory TNF alpha cytokine enhances the proliferation of human endometrial glandular cells. Obstet. Gynecol. Int. 2013, 368543.

Norheim, F., Hjorth, M., Langleite, T.M., Lee, S., Holen, T., Bindesboll, C., Stadheim, H.K., Gulseth, H.L., Birkeland, K.I., Kielland, A., Jensen, J., Dalen, K.T., Drevon, C.A., 2014. Regulation of angiopoietin-like protein 4 production during and after exercise. Physiol. Rep. 2.

O'Leary, V.B., Marchetti, C.M., Krishnan, R.K., Stetzer, B.P., Gonzalez, F., Kirwan, J.P., 2006. Exercise-induced reversal of insulin resistance in obese elderly is associated with reduced visceral fat. J. Appl. Physiol. (1985) 100, 1584–1589.

Ogawa, K., Sanada, K., Machida, S., Okutsu, M., Suzuki, K., 2010. Resistance exercise training-induced muscle hypertrophy was associated with reduction of inflammatory markers in elderly women. Mediators Inflamm. 2010, 171023.

Ogura, Y., Ouchi, N., Ohashi, K., Shibata, R., Kataoka, Y., Kambara, T., Kito, T., Maruyama, S., Yuasa, D., Matsuo, K., Enomoto, T., Uemura, Y., Miyabe, M., Ishii, M., Yamamoto, T., Shimizu, Y., Walsh, K., Murohara, T., 2012. Therapeutic impact of follistatin-like 1 on myocardial ischemic injury in preclinical models. Circulation 126, 1728–1738.

Olive, J.L., Miller, G.D., 2001. Differential effects of maximal- and moderate-intensity runs on plasma leptin in healthy trained subjects. Nutrition 17, 365–369.

Palacios-Ortega, S., Varela-Guruceaga, M., Algarabel, M., Ignacio Milagro, F., Alfredo Martinez, J., de Miguel, C., 2015. Effect of TNF-alpha on Caveolin-1 expression and insulin signaling during adipocyte differentiation and in mature adipocytes. Cell. Physiol. Biochem. 36, 1499−1516.

Park, H.K., Ahima, R.S., 2015. Physiology of leptin: energy homeostasis, neuroendocrine function and metabolism. Metabolism 64, 24−34.

Paulsen, G., Mikkelsen, U.R., Raastad, T., Peake, J.M., 2012. Leucocytes, cytokines and satellite cells: what role do they play in muscle damage and regeneration following eccentric exercise? Exerc. Immunol. Rev. 18, 42−97.

Peake, J.M., Suzuki, K., Hordern, M., Wilson, G., Nosaka, K., Coombes, J.S., 2005. Plasma cytokine changes in relation to exercise intensity and muscle damage. Eur. J. Appl. Physiol. 95, 514−521.

Pedersen, B.K., Febbraio, M., 2005. Muscle-derived interleukin-6−a possible link between skeletal muscle, adipose tissue, liver, and brain. Brain Behav. Immun. 19, 371−376.

Pedersen, B.K., Febbraio, M.A., 2008. Muscle as an endocrine organ: focus on muscle-derived interleukin-6. Physiol. Rev. 88, 1379−1406.

Pedersen, B.K., Febbraio, M.A., 2012. Muscles, exercise and obesity: skeletal muscle as a secretory organ. Nat. Rev. Endocrinol. 8, 457−465.

Pedersen, B.K., Steensberg, A., Fischer, C., Keller, C., Keller, P., Plomgaard, P., Febbraio, M., Saltin, B., 2003a. Searching for the exercise factor: is IL-6 a candidate? J. Muscle Res. Cell Motil. 24, 113−119.

Pedersen, B.K., Steensberg, A., Keller, P., Keller, C., Fischer, C., Hiscock, N., van Hall, G., Plomgaard, P., Febbraio, M.A., 2003b. Muscle-derived interleukin-6: lipolytic, anti-inflammatory and immune regulatory effects. Pflugers Arch. 446, 9−16.

Pedersen, B.K., Steensberg, A., Fischer, C., Keller, C., Keller, P., Plomgaard, P., Wolsk-Petersen, E., Febbraio, M., 2004. The metabolic role of IL-6 produced during exercise: is IL-6 an exercise factor? Proc. Nutr. Soc. 63, 263−267.

Perusse, L., Collier, G., Gagnon, J., Leon, A.S., Rao, D.C., Skinner, J.S., Wilmore, J.H., Nadeau, A., Zimmet, P.Z., Bouchard, C., 1997. Acute and chronic effects of exercise on leptin levels in humans. J. Appl. Physiol. 83, 5−10.

Petriz, B.A., Gomes, C.P., Almeida, J.A., de Oliveira Jr., G.P., Ribeiro, F.M., Pereira, R.W., Franco, O.L., 2017. The effects of acute and chronic exercise on skeletal muscle proteome. J. Cell. Physiol. 232, 257−269.

Phillips, A., Cobbold, C., 2014. A comparison of the effects of aerobic and intense exercise on the type 2 diabetes mellitus risk marker adipokines, adiponectin and retinol binding Protein-4. Int. J. Chronic. Dis. 2014, 358058.

Phillips, M.D., Patrizi, R.M., Cheek, D.J., Wooten, J.S., Barbee, J.J., Mitchell, J.B., 2012. Resistance training reduces subclinical inflammation in obese, postmenopausal women. Med. Sci. Sports Exerc. 44, 2099−2110.

Plomgaard, P., Nielsen, A.R., Fischer, C.P., Mortensen, O.H., Broholm, C., Penkowa, M., Krogh-Madsen, R., Erikstrup, C., Lindegaard, B., Petersen, A.M., Taudorf, S., Pedersen, B.K., 2007. Associations between insulin resistance and TNF-alpha in plasma, skeletal muscle and adipose tissue in humans with and without type 2 diabetes. Diabetologia 50, 2562−2571.

Polak, J., Klimcakova, E., Moro, C., Viguerie, N., Berlan, M., Hejnova, J., Richterova, B., Kraus, I., Langin, D., Stich, V., 2006. Effect of aerobic training on plasma levels and subcutaneous abdominal adipose tissue gene expression of adiponectin, leptin, interleukin 6, and tumor necrosis factor alpha in obese women. Metabolism 55, 1375−1381.

Qin, S., LaRosa, G., Campbell, J.J., Smith-Heath, H., Kassam, N., Shi, X., Zeng, L., Buthcher, E.C., Mackay, C.R., 1996. Expression of monocyte chemoattractant protein-1 and interleukin-8 receptors on subsets of T cells: correlation with transendothelial chemotactic potential. Eur. J. Immunol. 26, 640−647.

Rasmussen, P., Brassard, P., Adser, H., Pedersen, M.V., Leick, L., Hart, E., Secher, N.H., Pedersen, B.K., Pilegaard, H., 2009. Evidence for a release of brain-derived neurotrophic factor from the brain during exercise. Exp. Physiol. 94, 1062−1069.

Rauch, F., Bailey, D.A., Baxter-Jones, A., Mirwald, R., Faulkner, R., 2004. The 'muscle-bone unit' during the pubertal growth spurt. Bone 34, 771−775.

Rokling-Andersen, M.H., Reseland, J.E., Veierod, M.B., Anderssen, S.A., Jacobs Jr., D.R., Urdal, P., Jansson, J.O., Drevon, C.A., 2007. Effects of long-term exercise and diet intervention on plasma adipokine concentrations. Am. J. Clin. Nutr. 86, 1293−1301.

Rose-John, S., Winthrop, K., Calabrese, L., 2017. The role of IL-6 in host defence against infections: immunobiology and clinical implications. Nat. Rev. Rheumatol. 13, 399−409.

Roumaud, P., Martin, L.J., 2015. Roles of leptin, adiponectin and resistin in the transcriptional regulation of steroidogenic genes contributing to decreased Leydig cells function in obesity. Horm. Mol. Biol. Clin. Investig. 24, 25−45.

Sahibzada, H.A., Khurshid, Z., Khan, R.S., Naseem, M., Siddique, K.M., Mali, M., Zafar, M.S., 2017. Salivary IL-8, IL-6 and TNF-alpha as potential diagnostic biomarkers for oral cancer. Diagnostics 7.

Sakurai, T., Ogasawara, J., Kizaki, T., Sato, S., Ishibashi, Y., Takahashi, M., Kobayashi, O., Oh-Ishi, S., Nagasawa, J., Takahashi, K., Ishida, H., Izawa, T., Ohno, H., 2013. The effects of exercise training on obesity-induced dysregulated expression of adipokines in white adipose tissue. Int. J. Endocrinol. 2013, 801743.

Sakurai, T., Ogasawara, J., Shirato, K., Izawa, T., Oh-Ishi, S., Ishibashi, Y., Radak, Z., Ohno, H., Kizaki, T., 2017. Exercise training attenuates the dysregulated expression of adipokines and oxidative stress in white adipose tissue. Oxid. Med. Cell. Longev. 2017, 9410954.

Santa Mina, D., Connor, M.K., Alibhai, S.M., Toren, P., Guglietti, C., Matthew, A.G., Trachtenberg, J., Ritvo, P., 2013. Exercise effects on adipokines and the IGF axis in men with prostate cancer treated with androgen deprivation: a randomized study. Can. Urol. Assoc. J. 7, E692−E698.

Schmidt-Kassow, M., Schadle, S., Otterbein, S., Thiel, C., Doehring, A., Lotsch, J., Kaiser, J., 2012. Kinetics of serum brain-derived neurotrophic factor following low-intensity versus high-intensity exercise in men and women. Neuroreport 23, 889−893.

Serrano, A.L., Baeza-Raja, B., Perdiguero, E., Jardi, M., Munoz-Canoves, P., 2008. Interleukin-6 is an essential regulator of satellite cell-mediated skeletal muscle hypertrophy. Cell Metab. 7, 33−44.

Shi, Y., Massague, J., 2003. Mechanisms of TGF-beta signaling from cell membrane to the nucleus. Cell 113, 685−700.

Shinoda, Y., Yamaguchi, M., Ogata, N., Akune, T., Kubota, N., Yamauchi, T., Terauchi, Y., Kadowaki, T., Takeuchi, Y., Fukumoto, S., Ikeda, T., Hoshi, K., Chung, U.I., Nakamura, K., Kawaguchi, H., 2006. Regulation of bone formation by adiponectin through autocrine/paracrine and endocrine pathways. J. Cell. Biochem. 99, 196−208.

Tagliaferri, C., Wittrant, Y., Davicco, M.J., Walrand, S., Coxam, V., 2015. Muscle and bone, two interconnected tissues. Ageing Res. Rev. 21, 55−70.

Tantiwong, P., Shanmugasundaram, K., Monroy, A., Ghosh, S., Li, M., DeFronzo, R.A., Cersosimo, E., Sriwijitkamol, A., Mohan, S., Musi, N., 2010. NF-kappaB activity in muscle from obese and type 2 diabetic subjects under basal and exercise-stimulated conditions. Am. J. Physiol. Endocrinol. Metab. 299, E794–E801.

Thommesen, L., Stunes, A.K., Monjo, M., Grosvik, K., Tamburstuen, M.V., Kjobli, E., Lyngstadaas, S.P., Reseland, J.E., Syversen, U., 2006. Expression and regulation of resistin in osteoblasts and osteoclasts indicate a role in bone metabolism. J. Cell. Biochem. 99, 824–834.

Trayhurn, P., Drevon, C.A., Eckel, J., 2011. Secreted proteins from adipose tissue and skeletal muscle - adipokines, myokines and adipose/muscle cross-talk. Arch. Physiol. Biochem. 117, 47–56.

van Gemert, W.A., May, A.M., Schuit, A.J., Oosterhof, B.Y., Peeters, P.H., Monninkhof, E.M., 2016. Effect of weight loss with or without exercise on inflammatory markers and adipokines in postmenopausal women: the SHAPE-2 trial, a randomized controlled trial. Cancer Epidemiol. Biomarkers Prev. 25, 799–806.

Varady, K.A., Bhutani, S., Church, E.C., Phillips, S.A., 2010. Adipokine responses to acute resistance exercise in trained and untrained men. Med. Sci. Sports Exerc. 42, 456–462.

Vella, L., Caldow, M.K., Larsen, A.E., Tassoni, D., Della Gatta, P.A., Gran, P., Russell, A.P., Cameron-Smith, D., 2012. Resistance exercise increases NF-kappaB activity in human skeletal muscle. Am. J. Physiol. Regul. Integr. Comp. Physiol. 302, R667–R673.

Walker, R.G., Poggioli, T., Katsimpardi, L., Buchanan, S.M., Oh, J., Wattrus, S., Heidecker, B., Fong, Y.W., Rubin, L.L., Ganz, P., Thompson, T.B., Wagers, A.J., Lee, R.T., 2016. Biochemistry and Biology of GDF11 and myostatin: similarities, differences, and questions for future investigation. Circ. Res. 118, 1125–1141 discussion 42.

Wolsk, E., Mygind, H., Grondahl, T.S., Pedersen, B.K., van Hall, G., 2012. Human skeletal muscle releases leptin in vivo. Cytokine 60, 667–673.

Yang, C.B., Chuang, C.C., Kuo, C.S., Hsu, C.H., Tsao, T.H., 2014. Effects of an acute bout of exercise on serum soluble leptin receptor (sOB-R) levels. J. Sports Sci. 32, 446–451.

Yu, N., Ruan, Y., Gao, X., Sun, J., 2017. Systematic review and meta-analysis of randomized, controlled trials on the effect of exercise on serum leptin and adiponectin in overweight and obese individuals. Horm. Metab. Res. 49, 164–173.

Secretory Malfunction: A Key Step to Metabolic Diseases

CHAPTER OUTLINE

In the previous chapters, attempts have been made to provide a comprehensive view on the secretory function of adipose tissue and skeletal muscle under physiological conditions. The crosstalk between these two tissues involves a host of molecules and definitely extends to multiple other organs, including the liver, heart, brain, pancreas, and vasculature (Li et al., 2017; Zhang et al., 2017; Wang and Yang, 2017; Shirakawa et al., 2017; Rodriguez et al., 2017; Di Raimondo et al., 2016; Stern et al., 2016). At present, we are still far away from understanding the fine-tuning of this complex network that provides a highly sensitive machinery that orchestrates numerous metabolic functions and most likely is a key determinant of metabolic homeostasis. Given this, it is conceivable that an aberrant secretory process, specifically in adipose tissue as a major endocrine crosstalk agent, will lead to perturbations of metabolic homeostasis, finally leading to the manifestation of metabolic diseases.

This consideration highlights the importance of the cellular secretome and makes it an important target for novel therapeutic approaches. At present, this is hampered by the enormous complexity of the secretome and our still-limited knowledge regarding the crosstalk scenario.

Secretory malfunction has been extensively investigated in adipose tissue, given that obesity, which is defined as the expansion of fat, has been recognized as a major risk factor for metabolic and related diseases such as type 2 diabetes (T2D), fatty liver disease, cardiovascular (CV) disorders, diseases of the central nervous system, obstructive sleep apnea, and different types of cancer (Fruh, 2017; Goodarzi, 2017; Cuthbertson et al., 2017). It is now generally thought that adipose tissue inflammation, which is a type of chronic, subacute inflammation, is a key driver of the secretory dysfunction leading to a change of the adipokine and adipocytokine pattern. We will therefore initially focus on the mechanisms of adipose tissue inflammation and the impact for insulin resistance and T2D. This mostly relates to the negative crosstalk between adipose tissue and skeletal muscle and can be considered as a paradigm of secretory dysfunction leading to metabolic complications. Another very important topic relates to adipose tissue inflammation and cardiovascular disease (CVD), which we will consider in depth.

In addition to adipose tissue, lipid accumulation in the liver (nonalcoholic fatty liver disease, NAFLD) leads to disturbed liver function and also plays a major role in the pathogenesis of T2D (Radaelli et al., 2017; Alisi et al., 2017; Tilg et al., 2017; Younossi et al., 2017). This implies changes of the liver secretome, a more recently studied topic. Finally, a sedentary lifestyle substantially affects the muscle secretome, and we will analyze this scenario in more detail (Gonzalez et al., 2017a).

ADIPOSE TISSUE INFLAMMATION

Enlargement of adipose tissue ensues as a result of an imbalance between energy intake and expenditure. As a consequence, adipocytes partly undergo hyperplasia but mostly hypertrophy to meet the increased demand for storage capacities (Hirsch and Batchelor, 1976). Hypertrophy of adipocytes implies a chronic cellular stress situation and represents a key event related to the loss of insulin sensitivity in obese subjects (Cotillard et al., 2014). A persistent state of energy excess represents an increased burden for the lipid storage and processing capacities of the expanding adipose tissue, resulting in various dysfunctions within the tissue such as low-grade chronic inflammation and hypoxia (Karalis et al., 2009; Wood et al., 2009). These obesity-associated dysfunctions may lead to changes in the cellular composition of the tissue, including alterations in the number, phenotype, and localization of immune, vascular, and structural cells, finally resulting in an altered adipose tissue secretome and an altered crosstalk scenario.

Evidence supports the notion that hypoxia may occur in areas within adipose tissue in obesity as a result of adipocyte hypertrophy, compromising effective O_2 supply from the vasculature, thereby inducing an inflammatory response through recruitment of the transcription factor hypoxia-inducible factor-1 (Engin, 2017; Rasouli, 2016; Ryan, 2017; Trayhurn, 2013, 2014; Trayhurn, Wang, and Wood, 2008a, 2008b; Wood et al., 2009; Ye, 2009). Furthermore, studies in animal models (mutant mice, diet-induced obesity) and cell culture systems (mouse and human adipocytes) have provided strong support for a role for hypoxia in altering the production of several inflammation-related adipokines, such as IL-6, leptin, and many others. Increased glucose transport into adipocytes is also observed with low O_2 tension, largely as a result of the upregulation of GLUT-1 expression, indicating changes in cellular glucose metabolism. It is also important to note that hypoxia induces inflammatory responses in macrophages and inhibits adipogenesis by reducing the differentiation of preadipocytes. Overall, the work performed by the Trayhurn lab (see references mentioned earlier) provides strong evidence to assume that cellular hypoxia may be a key factor in adipocyte physiology and the underlying cause of adipose tissue dysfunction contributing to the adverse metabolic milieu associated with obesity.

Inflammation is a highly complex biological process represented by a series of cellular and molecular responses that function to defend the organism from infections or other insults (Vane and Botting, 1987; Liu and Malik, 2006; Heymann, 2006; Cronstein, 1992; Busse et al., 1993). Dysregulated inflammation is widely considered as a major cause of multiple diseases including atherosclerosis, asthma, and other autoimmune diseases. The inflammation process is under the control of a variety of immune cells, including neutrophils, macrophages, dendritic cells, mast cells, and different groups of lymphocytes. A detailed description of the different steps of inflammation is beyond the scope of this chapter, and the reader should consider the review articles cited previously. Briefly, neutrophils are the first to migrate into the site of infection, where they produce a number of chemokines that are able and required to recruit a series of other immune cells, with macrophages first entering the site of infection, followed by the recruitment of lymphocytes. Thus, in a nutshell, inflammation involves increased cytokine levels in the circulation and in local sites, concomitant with an increased infiltration of immune cells. It is important to note that adipose tissue inflammation is a so-called low-grade, chronic inflammation, specifically characterized by a much lower level of circulating cytokines and preferential infiltration of macrophages into the inflamed fat cell area (Weisberg et al., 2003; Ferrante, 2007). The current view of adipose tissue inflammation is that obesity increases the number of macrophages in adipose tissue and that these cells are responsible for the production of cytokines in obesity. Furthermore, many other types of immune cells were found in adipose tissue, and these cells play an additional role in the regulation of obesity-associated inflammation (Ferrante, 2007).

Macrophages are considered to play a key role in the regulation of chronic inflammation. Basically, macrophages are a rather diverse group of cells with substantial differences between different tissues. Importantly, subpopulations of macrophages exist in the same tissue that can be (in a simplified system) considered as M1 and M2 macrophages (Gordon, 2003). M1 macrophages are considered to be proinflammatory and express IL-6 and tumor necrosis factor α (TNFα). M2 macrophages play a role in antiinflammatory pathways and wound healing.

Increased expression of chemoattractant proteins such as monocyte chemotactic protein 1 (MCP-1) induces recruitment and infiltration of additional macrophages. These contribute to the increased expression of proinflammatory cytokines such as TNFα, thereby further exacerbating the obesity-associated inflamed status of the adipose tissue (Xu et al., 2003). Additionally, obesity-associated adipocyte hypertrophy has also been associated with a shift of the adipocyte secretome to a more proinflammatory composition (Skurk et al., 2007). In this context, a positive correlation has been described between adipocyte size and secretion of various proinflammatory factors such as TNFα, IL-6, IL-8, MCP-1, leptin, and granulocyte colony stimulating factor.

These observations demonstrate that the adipose tissue output dramatically changes in pathological conditions such as obesity. However, the adipokinome may already vary depending on the site of the different adipose tissue depots (Ouchi et al., 2011). Adipose tissue depots in the visceral and the subcutaneous compartment are the two most abundant depots, and it has been shown that they produce unique profiles of adipokines. In this context, visceral adipose tissue has received special attention because various studies have found a positive correlation between the amount of visceral adipose tissue and CVDs (Bosello and Zamboni, 2000). Recent studies have even proposed that visceral adiposity, measured as waist circumference, may represent a more precise risk indicator for T2D and CVDs than whole-body obesity (Lofgren et al., 2004). One reason for a specific pathophysiological role of the visceral depot may be attributed to its location, as it drains directly into the portal vein. Furthermore, the visceral adipokinome contains many proinflammatory risk factors, which contribute to the close association of visceral adipose tissue and obesity-associated complications (Matsuzawa et al. 1994, 1995, 2011).

OBESITY, INSULIN RESISTANCE, AND TYPE 2 DIABETES

Obesity is now considered as an epidemic disease decreasing life expectancy owing to adverse consequences such as T2D (Haslam and James, 2005; Verma and Hussain, 2017; Czech, 2017). In fact, about 80% of diabetes cases can be attributed to weight gain (IDF, 2003). Obesity strongly predisposes individuals to the development of diabetes, as a very high body mass index (BMI) (>40) is associated with a near 100% risk for the disease (Anderson et al., 2003). Furthermore, there is undoubtedly a strong causal link between increased adipose tissue mass in obesity and insulin resistance in peripheral tissues such as liver and skeletal muscle in diabetic patients (Argiles et al., 2005; Krebs and Roden, 2005; Yazici and Sezer, 2017; Wu and Ballantyne, 2017; Yamaguchi and Yoshino, 2017). Insulin resistance is defined as a condition where an organism does not respond appropriately to circulating insulin. In addition to predisposing for the development of T2D, insulin resistance is also associated with CVD, a topic discussed in the next subchapter. Insulin resistance is not limited to the classical target tissues of insulin, namely adipose tissue, liver, and skeletal muscle, but can also be found in other tissues such as vascular cells, the brain, and the pancreas. Most notably, skeletal muscle from obese patients displays insulin resistance in vivo (Brozinick et al., 2003), a feature that might be acquired with obesity, as cultured skeletal muscle cells from obese patients with insulin resistance do not exhibit altered insulin responsiveness (Pender et al., 2005). Induction of skeletal muscle insulin resistance due to secretory malfunction of adipose tissue is a paradigm of negative crosstalk between the two organs and has gained considerable interest as a molecular link to a number of metabolic diseases. The direct crosstalk between human adipocytes and myocytes was first demonstrated in a coculture model where both cell types share the same medium (Dietze et al., 2002). Coculture with adipocytes clearly leads to insulin resistance in skeletal muscle cells, as the latter display decreased insulin receptor substrate (IRS)-1 and Akt phosphorylation with subsequent inhibition of glucose transporter-4 translocation and glucose uptake (Dietze et al., 2002; Dietze-Schroeder et al., 2005). Although simplified, this model clearly reflects critical features of insulin resistance observed in skeletal muscle from diabetic patients (Zierath et al., 2000). Analysis of the secretion from primary human adipocytes revealed that these cells release classical adipokines such as TNFα, IL-6, leptin, and adiponectin and newly discovered adipokines such as MCP-1, tissue inhibitor of metalloproteinases-1, and retinol-binding protein-4 (Sartipy and Loskutoff, 2003; Yang et al., 2005).

The increased expression and secretion of several adipokines in the obese state is considered as an indication of low-grade chronic inflammation in adipose tissue. Protein kinase C (PKC) and IKK are two kinases that were shown to play an important role in inflammatory processes underlying insulin resistance, particularly with regard to lipotoxicity (Wellen and Hotamisligil, 2005). Various PKC isoforms have been shown to be involved in lipid-induced insulin resistance in skeletal muscle in vitro (Haasch et al., 2006) and in vivo (Bandyopadhyay et al., 2005). IKK influences insulin sensitivity, especially in skeletal muscle, through two different mechanisms. First, it can directly inhibit insulin signaling by phosphorylating IRS-1 on serine residues (Gao et al., 2002), and second, IKK activates nuclear factor (NF)-κB. In turn, NF-κB regulates production of proinflammatory cytokines such as TNFα and IL-6 (Shoelson et al., 2003). Accordingly, inhibition of IKK reduces the secretion of proinflammatory adipokines and prevents insulin resistance in vitro and in vivo (Dietze et al., 2004; Arkan et al., 2005; Lappas et al., 2005).

Endoplasmic reticulum (ER) stress is another mechanism that could be involved in insulin resistance resulting from crosstalk between increased adipose tissue mass and skeletal muscle. ER stress, indicated by hyperactivation of c-Jun N-terminal kinase (JNK) and subsequent inhibitory phosphorylation of IRS-1, can be observed in adipose tissue and liver of *ob/ob* obese mice and high-fat diet—induced obese mice but not in skeletal muscle (Ozcan et al., 2004). The "unfolded protein response" and ER chaperones are part of ER stress and seem to be altered in diabetes (Zhao and Ackerman, 2006). ER stress seems to play no role in skeletal muscle even if JNK activation in skeletal muscle from obese and diabetic patients can be observed (Bandyopadhyay et al., 2005). Therefore, other mechanisms distinct from ER stress must lead to JNK activation in skeletal muscle possibly involving Toll-like receptor 4 (TLR4) and the NF-κB inflammatory pathway (Frost et al., 2004).

Defective oxidative phosphorylation is another mechanism that seems to be involved in insulin resistance in both skeletal muscle (Mootha et al., 2003) and adipose tissue (Dahlman et al., 2006). Several genes involved in defense against oxidative stress are defective in diabetic patients (Bruce et al., 2003) and correlate with muscle oxidative capacity. Furthermore, mitochondrial activity is already decreased in insulin-resistant relatives of diabetic patients who are at high risk to become diabetic later in life (Petersen et al., 2004). This loss of mitochondrial oxidative capacity can be explained by the fact that insulin resistance is accompanied by fewer muscle mitochondria. Accordingly, skeletal muscle from obese patients also contains a smaller number of mitochondria owing to a lower ratio of type 1 to type 2 muscle fibers. Furthermore, a deficit in muscle oxidative phosphorylation could also be caused by defects in mitochondrial function. In fact, analysis of gene expression in skeletal muscle of diabetic patients revealed lower expression of a cluster of oxidative genes related to peroxisome proliferator–activated receptor γ (PPARγ) coactivator 1 (Mootha et al., 2003) and mitochondrial dysfunction (Kelley et al., 2002). Whether intramyocellular lipid accumulation is the cause or an effect of decreased mitochondrial oxidative capacity is somewhat unclear (Schrauwen and Hesselink, 2004).

More recent evidence supports the view that the so-called inflammasome plays an important role in the induction of obesity and insulin resistance (Wen et al., 2011; Vandanmagsar et al., 2011; Stienstra et al., 2011). The inflammasome is an intracellular multimeric complex of different proteins and is a key player in innate immunity (Heilig and Broz, 2017; Toldo and Abbate, 2017; Place and Kanneganti, 2017; Baldrighi et al., 2017; Kim et al., 2017; Kopitar-Jerala, 2017). The inflammasome is of key importance to induce the maturation of IL-1β and IL-18 and to produce these proinflammatory cytokines in response to different danger signals. This process involves the increased transcription of pro-IL-1β and pro-IL-18 and the activation of the inflammasome. The involvement of the inflammasome in obesity-induced insulin resistance was supported by the deletion of inflammasome components. This resulted in suppression of inflammation and improvement of insulin resistance. In this context, the role of IL-1β gained considerable interest. In vitro treatment of adipocytes with IL-1β showed an induction of insulin resistance (Jager et al., 2007). These authors found that prolonged IL-1β treatment reduced the insulin-induced glucose uptake, whereas an acute treatment had no effect. Chronic treatment with IL-1β slightly decreased the expression of GLUT 4 and markedly inhibited its translocation to the plasma membrane in response to insulin. This inhibitory effect was due to a decrease in the amount of IRS-1 but not IRS-2 expression in both 3T3-L1 and human adipocytes. The decrease in IRS-1 amount resulted in a reduction in its tyrosine phosphorylation and the alteration of insulin-induced PKB activation and AS160 phosphorylation. Pharmacological inhibition of ERK totally inhibited IL-1β—induced downregulation of IRS-1 mRNA. It was concluded that IL-1β reduces IRS-1 expression at a transcriptional level through a mechanism that is ERK-dependent and at a posttranscriptional level independently of ERK activation. By targeting IRS-1, IL-1β is capable of impairing insulin signaling and action, and could thus participate in concert with other cytokines, in the development of insulin resistance in adipocytes. Thus, IL-1β can be considered as a paradigm of a negative mediator of insulin action, specifically under conditions of obesity-induced inflammation and secretory malfunction.

The in vitro data on IL-1β were further substantiated by in vivo work when obese animals were treated with a neutralizing antibody against IL-1β, resulting in an improvement of insulin resistance (Osborn et al., 2008). It was reported that after 13 weeks of treatment the IL-1β antibody—treated group showed reduced glycated hemoglobin, reduced serum levels of proinsulin, reduced levels of insulin, and smaller islet size relative to the control antibody—treated group. Although there was no improvement of obesity, a significant improvement of glycemic control and of β-cell function was achieved by this pharmacological treatment, which may slow/prevent disease progression in T2D. In a seminal clinical trial it was then reported that anakinra, an IL-1 receptor antagonist, improves glycemic control in type 2 diabetic subjects by improving β-cell function (Larsen et al., 2007). Treatment with anakinra also lowered the level of systemic inflammation markers and suggested that inhibition of inflammasome activity improves insulin resistance and T2D in humans. However, specific inactivation of IL-1β by canakinumab, a monoclonal antibody against IL-1β, failed to improve glycemic control in diabetic subjects despite lower inflammation (Ridker et al., 2012). Thus, the precise role of the inflammasome in obesity-induced insulin resistance and T2D needs to be further explored.

Because obesity-associated inflammation is coupled to the infiltration of immune cells, it is generally thought that the production of cytokines by these cells represents the crosstalk signal to the insulin-responsive cells such as myocytes, adipocytes, and others. However, activation of inflammatory pathways in immune cells is known to lead to the production of both pro- and antiinflammatory pathways. This scenario has complicated the process of identifying single cytokines as mediators of insulin resistance, and controversial data exist from extensive in vitro and in vivo studies using genetically modified mouse models. A classical example is TNFα, which is known to compromise insulin signaling, most likely by activating downstream kinases that lead to serine phosphorylation of IRS-1, with the consequence of reduced insulin signaling (Rui et al., 2001). Subsequent clinical studies using an anti-TNFα antibody showed that inactivation of TNFα does not improve glycemic control in type 2 diabetic subjects (Paquot et al., 2000; Ofei et al., 1996). Divergent results were also obtained for IL-6, a well-known proinflammatory cytokine and an established risk factor for the development of T2D in humans. Animal studies showed that infusion of IL-6 induces insulin resistance in the absence of obesity; however, deletion of IL-6 leads to the development of obesity and insulin resistance (Wallenius et al., 2002). Antiinflammatory cytokines such as IL-10 have been suggested to promote the improvement of obesity-induced insulin resistance. However, IL-10–knockout (KO) mouse data showed that this does not counteract diet-induced insulin resistance or inflammation (Kowalski et al., 2011). Thus, it can be concluded that the metabolic and inflammatory function of selected cytokines is divergent and may possibly reflect the temporal and local regulation of inflammation. More likely, it appears that a single cytokine is not sufficient to induce obesity-associated inflammation and insulin resistance, requiring a complex network of several cytokines to finally induce a disease-related dysfunction of insulin sensitivity in target tissues.

ADIPOSE TISSUE DYSFUNCTION AND CARDIOVASCULAR DISEASE

As pointed out before, obesity and adipose tissue inflammation represent a major risk factor for CVDs (Kachur et al., 2017; Ellulu, 2017; Gonzalez et al., 2017b). These disorders are characterized by a variety of vessel alterations, with atherosclerosis and vascular inflammation being the key event. This generated a deep interest in the adipovascular axis and the effects of various adipokines on cells of the vascular wall. In Chapter 2, we have already briefly discussed the physiological impact of the adipovascular axis, focusing on adiponectin, leptin, DPP4, and visfatin. Here we provide a more detailed analysis of the pathophysiological impact of the adipovascular axis and, additionally, address some novel players in this scenario such as omentin and apelin. Because augmented proliferation and migration of smooth muscle cells represents a central event in atherosclerosis, adipokines able to induce these processes have been studied in detail (Ruscica et al., 2017; Fulop et al., 2016). Furthermore, the activation of proinflammatory signaling pathways by adipocytokines such as TNFα, leptin, and many others is considered to contribute significantly to the development of CVD (Sattin and Towheed, 2016). In this context, the transcription factor NF-κB plays a dominant role because its downstream target genes represent adhesion molecules, proinflammatory cytokines, and mediators of proliferation. In 2012, our group published the first detailed and comprehensive analysis of the adipovascular axis and identified more than 30 adipokines being associated with CVD (Taube et al., 2012). The most prominent link was found for leptin, TNFα, interleukins, and several novel adipokines. These molecules are of great interest as potential drug targets and biomarkers.

DPP4 AND DPP4 INHIBITORS

By using a comprehensive proteomic profiling approach of the human adipocyte secretome, we identified DPP4 as a novel adipokine (Lamers et al., 2011). DPP4 is a ubiquitously expressed cell surface protease, which gained considerable attention for its role in the regulation of glycemia through cleavage and inactivation of incretin hormones. Owing to a process called shedding, a soluble form of DPP4 (sDPP4) lacking the cytoplasmatic tail and the transmembrane region is released from the cell surface and can be found in plasma and other body fluids (Rohrborn et al., 2015). Interestingly, the serum concentrations of DPP4 positively correlate with various parameters of the metabolic syndrome such as BMI, adipocyte surface, and leptin and insulin levels (Lamers et al., 2011) and are increased in inflammatory disease and atherosclerosis. A comparison of different fat depots demonstrated the highest DPP4 expression in visceral adipose tissue of obese patients (Sell et al., 2013). The development of DPP4 inhibitors and a central role for these agents in glycemic control emphasized the importance of DPP4 and its potential role in CV physiology and pathology. The functional role of DPP4 is related to both its enzymatic and nonenzymatic function. DPP4 is a serine exopeptidase, which cleaves X-proline dipeptides from the N-terminus of several substrates, including chemokines, neuropeptides, and vasoactive peptides (Rohrborn et al., 2015). DPP4 is not only an enzyme but may also interact with a range of ligands, including ADA, caveolin-1, thromboxane A2 receptor, and fibronectin (Zhong et al., 2013). The binding of DPP4 to these ligands plays a role in various processes including immune regulatory function. Recent studies indicated that in addition to regulation of postprandial glycemia, DPP4 may exert several functions on the CV system, both directly and indirectly by its inhibition.

DPP4 INHIBITORS: INDIRECT EFFECTS OF DPP4

Because DPP4 inhibitors are available drugs approved for the treatment of T2D, extensive research exists focusing on the effects of DPP4 inhibition with only limited information regarding direct effects of sDPP4. Inhibitors of DPP4 are able to prolong the bioavailability of the endogenously secreted incretin hormone GLP-1 and the glucose-dependent insulinotropic polypeptide. More recent evidence indicates that DPP4 inhibitors also have important protective effects on the CV system, through endothelial repair, antiinflammatory effects, and blunting of ischemic injury (Fadini and Avogaro, 2011). It is established that GLP-1 itself has favorable CV effects; therefore, most of the beneficial effects of DPP4 inhibitors are ascribed to increased GLP-1 bioavailability and signaling. GLP-1 analogues such as exendin-4 stimulate proliferation of human coronary artery endothelial cells through endothelial nitric oxide synthase (eNOS)−, PKA− and phosphoinositol-3 kinase (PI3K/Akt)−dependent pathways. It was also reported that continuous infusion of exendin-4 reduces neointimal formation at 4 weeks after injury without affecting body weight or various metabolic parameters. As a clinical readout, a large retrospective analysis reported that patients prescribed with the GLP-1 analogue exenatide had a significant 20% reduction of CVD events compared with patients on other glucose-lowering agents. It has to be taken into account that these studies were carried out using either native GLP-1 or recombinant GLP-1 analogues at high concentrations or in a way that induced supraphysiological GLP-1 signaling. Considering that DPP4 inhibition restores GLP-1 signaling within the physiological range, beneficial effects of DPP4 inhibitors might be different to those of GLP-1 analogues.

In addition to GLP-1, alternative substrates of DPP4 do exist, which might play a role in the favorable CV effects of DPP4 inhibitors. Two of the most promising candidates are stromal-derived factor (SDF)-1α and neuropeptide Y (NPY). SDF-1α is a key regulator of endothelial progenitor cells (EPCs), which acts by binding to their receptor CXCR4. EPCs are derived from the bone marrow and can be mobilized into the circulation in response to a host of different stimuli. EPCs adhere to the endothelium at sites of hypoxia or vascular damage and are able to participate in new vessel formation. Animal experiments demonstrated that genetic deletion or pharmacological inhibition of DPP4 is able to increase the homing of CXCR4$^+$ EPCs at sites of myocardial damage, resulting in a reduced cardiac remodeling and improved heart function and survival (Fadini and Avogaro, 2011). Also in a human study Fadini et al. could demonstrate that type 2 diabetic patients receiving a 4-week course of therapy with the DPP4 inhibitor sitagliptin show increased SDF-1α plasma concentrations and circulating EPC levels. These effects were unrelated to changes in nitrite/nitrate levels or to reduction in plasma glucose, suggesting that they are GLP-1— independent. These studies point to a potential new therapeutic strategy to use DPP4 inhibitors for vascular repair through stimulation of EPCs. The reported effects of DPP4 inhibition on angiogenesis could also be mediated by modulation of the NPY signaling pathway. DPP4 converts NPY1—36 to its shorter form (NPY3—36), thus shifting its activity from Y1-mediated vasoconstriction and vascular smooth muscle cell growth to Y2/Y5-mediated angiogenesis (Fadini and Avogaro, 2011).

Several studies exist suggesting that the direct effects of DPP4 inhibitors cannot be explained by an increased bioavailability of DPP4 substrates. Ta et al. reported that the specific DPP4 inhibitor alogliptin blocks lipopolysaccharide (LPS)-induced ERK phosphorylation in U937 histiocytes, representing a model of cells involved in the progression of atherosclerosis, such as foam cells (Ta et al., 2010). Downstream of ERK, the inhibition of DPP4 prevented the activation of matrix metalloproteinases (MMPs) via the TLR pathway, a process contributing to the destabilization of atherosclerotic plaques, finally leading to acute vascular events. It was also demonstrated that sitagliptin prevents TNFα-induced PAI-1, ICAM, and VCAM gene and protein expression in vascular endothelial cells independent of GLP-1 (Hu et al., 2013). This effect could partially be explained by a reduction of the TNFα-induced NF-κB mRNA expression by sitagliptin. Furthermore, incubation of human umbilical vein endothelial cells with the DPP4 inhibitor alogliptin resulted in eNOS and Akt phosphorylation (Ser1177 and Ser473, respectively) and a concomitant rapid increase in nitric oxide (Shah et al., 2011c). In agreement with these findings, alogliptin causes acute vascular relaxation in a nondiabetic mouse model through GLP-1—independent pathways that are both nitric oxide— and endothelium-derived hyperpolarizing factor—dependent. Overall, these results support the notion that DPP4 itself plays a role in the beneficial effects of DPP4 inhibitors.

DIRECT EFFECTS OF SOLUBLE DPP4

The direct effects of sDPP4 on vascular cells have only recently been investigated, and the available data on this topic are still limited. Our group was the first to show direct effects of sDPP4 on human vascular smooth muscle cells (hVSMCs), illustrated by increased proliferation and the induction of inflammation (Wronkowitz et al., 2014a). Using DPP4 concentrations that reflect serum levels of lean and obese patients, sDPP4 induced a marked activation of the MAPK signaling pathway, resulting in increased proliferation of hVSMCs (Wronkowitz et al., 2014a). It was also found that DPP4 directly activates the proinflammatory NF-κB signaling pathway, leading to an increased expression and secretion of cytokines such as IL-6, IL-8, and MCP-1 that exert a proinflammatory function. Because all DPP4-induced effects could be completely prevented by DPP4 inhibition, it was concluded that the direct effects of sDPP4 were dependent on its enzymatic activity. However, to identify the exact mechanisms how DPP4 exerts its signaling properties, further investigations will be required. Recently, direct effects of DPP4 on vascular function with special focus on vascular reactivity were also explored (Romacho et al., 2016). Neither the contractility to noradrenaline nor the endothelium-independent relaxations induced by sodium nitroprusside were modified by sDPP4. However, sDPP4 impaired in a concentration-dependent manner the endothelium-dependent relaxation elicited by acetylcholine. The DPP4 inhibitors K579 and linagliptin prevented the defective relaxation induced by sDPP4, as did the protease-activated receptor 2 (PAR2) inhibitor GB83. Downstream of PAR2, the cyclooxygenase (COX) inhibitor indomethacin, the COX2 inhibitor celecoxib, or the thromboxane receptor blocker SQ29548 prevented the deleterious effects of sDPP4. Accordingly, sDPP4 triggered the release of TXA2 by endothelial cells, whereas TXA2 release was prevented by inhibiting DPP4, PAR2, or COX. These data suggest that DPP4 directly impairs endothelium-dependent relaxation, through a mechanism that involves COX activation and likely the release of a vasoconstrictor prostanoid.

DPP4 is also well known to play a role in immunomodulation. Given that atherosclerosis is an immune inflammatory disease, it may be speculated that the inhibition of DPP4 modulates responses occurring within early or late atherosclerotic lesions. When using low-density lipoprotein receptor–deficient (LDLR−/−) mice, it was demonstrated that exogenously injected DPP4 increases monocyte migration in vivo (Shah et al., 2011b). These promigratory properties of DPP4 could be completely inhibited by the DPP4 inhibitor sitagliptin. Overall, these studies on direct effects of sDPP4 might contribute to explain the vascular protective effects of DPP4 inhibitors independent of GLP-1. The existing data strongly support the notion that DPP4 is a promising candidate linking adipose tissue dysfunction to CVD (Wronkowitz et al., 2014b).

LIPOCALIN-2 AND VISFATIN

LIPOCALIN-2

Lipocalin-2 (or neutrophil gelatinase–associated lipocalin) is a small lipid binding protein, but the ligands still await identification (Liu et al., 2012; Jang et al., 2012;

Wang, 2012). It is expressed in a variety of tissues including adipose tissue where it is found to be highly expressed in mature adipocytes compared with undifferentiated preadipocytes. In addition to lipocalin's well-characterized function in the innate immune response, some data suggest that it also plays a role as a proinflammatory adipokine in obesity and in obesity-associated metabolic diseases (Moschen et al., 2017).

Serum levels of lipocalin-2 are increased in the obese state and in patients with atherosclerosis, hypertension, coronary artery disease, and coronary heart disease (Sell et al., 2012). Augmented concentrations are predictive of increased mortality after myocardial infarction and correlate with severity of coronary heart disease, making it likely that lipocalin-2 could be an interesting biomarker. Experimental data from rodents have established that adipose tissue might be a source of lipocalin-2 in obesity. In addition, a high expression of lipocalin-2 can be found in atheromatous human plaques in association with increased MMP-9 activity. The precise role of lipocalin-2 in endothelial dysfunction preceding CVDs is still under investigation. Importantly, lipocalin-2–KO mice are protected from aging-associated and dietary obesity–associated endothelial dysfunctions (Liu et al., 2012). Owing to its role as a lipid carrier, lipocalin-2 might elevate endothelium-dependent vasoconstriction and attenuate relaxation by affecting eNOS activity and increased COX expression in intact arteries and primary endothelial cells. It was also shown that the expression of lipocalin-2 can be induced by various proinflammatory stimuli including IL-1β, TNFα, dexamethasone, and LPS. Lipocalin-2 itself induces inflammation in various cell types including cardiomyocytes. Treatment with recombinant lipocalin-2 induces cardiomyocyte apoptosis in parallel to macrophage infiltration in the myocardium. It has also been suggested that lipocalin-2 may mediate postischemic inflammatory and remodeling responses. Overall, lipocalin-2 is a mediator of inflammation and CVDs with a potential to be used as a biomarker and a therapeutic target in the context of obesity. Future studies are needed to fully decipher the various biological functions of this interesting adipokine.

VISFATIN

Visfatin was initially reported as an insulin-mimetic adipokine, which binds to and activates the insulin receptor, promotes adipogenesis, stimulates glucose uptake in vitro, and exerts glucose-lowering effects in mice in vivo (Fukuhara et al., 2005). However, this effect of visfatin could not be reproduced by other groups and was retracted in 2007 (Fukuhara et al., 2007). However, the participation of the insulin receptor in visfatin-mediated actions still remains unclear, and the insulin-mimetic properties of visfatin remain elusive (Revollo et al., 2007). The term visfatin refers to visceral fat because this depot was found to represent the main source of visfatin in both mice and humans. However, subsequent reports showed similar visfatin levels in human subcutaneous and visceral fat. Importantly, visfatin is also expressed in perivascular and epicardial fat where it might exert a paracrine CV action. As reported in a recent review on visfatin, activated macrophages in human visceral adipose tissue appear to release higher amounts of visfatin compared with adipocytes (Romacho et al., 2013).

In both human and murine adipocytes, visfatin expression and release are strongly upregulated during differentiation. Importantly, visfatin is ubiquitously expressed and exerts multiple biological actions different from the action profile of an adipokine. Thus, visfatin is identical to pre−B-cell colony-enhancing factor (PBEF), a cytokine known to promote maturation of early B-lineage precursor cells. Most importantly, it is now accepted that visfatin exhibits intrinsic enzymatic activity as a nicotinamide phosphoribosyltransferase (Nampt), catalyzing the rate-limiting step leading to the synthesis of nicotinamide adenine dinucleotide (NAD) (Romacho et al., 2013). Although some conflicting results do exist, a number of studies have reported elevated circulating levels of visfatin in obesity, T2D, and the metabolic syndrome. Moreover, different studies have established positive associations between enhanced serum visfatin levels and atherothrombotic diseases, suggesting that this protein may represent a biomarker of metabolic-related CV complications. In both patients with the metabolic syndrome and T2D, data suggested that the level of visfatin may serve as a marker of advanced carotid atherosclerosis for type 2 diabetic patients. In morbid obese patients, epicardial fat thickness correlates with enhanced visfatin and with visceral obesity. It is also important to note that visfatin levels are positively associated to circulating inflammatory markers, such as IL-6, C-reactive protein (CRP) and MCP-1. It is therefore possible that high levels of visfatin reflect changes in the systemic inflammation, independently of the insulin resistance state (Romacho et al., 2013). The potential pharmacological regulation of the circulating levels of visfatin is not fully understood, as there are discrepant in vivo and in vitro studies. Dexamethasone upregulates visfatin expression in vitro in 3T3-L1 adipocytes, although not affecting the serum levels in humans. When considering antidiabetic drugs, it was found that troglitazone suppresses visfatin expression in 3T3-L1 adipocytes, whereas rosiglitazone increases visfatin secretion in human adipocytes. Conflicting results on the effects of weight loss, exercise, and antiobesity drugs on visfatin levels do also exist, and future studies on this topic are required.

Increasing in vitro evidence points to a detrimental role for visfatin in the CV system. Based on these data, visfatin should be regarded as a proatherogenic factor promoting vascular cell proliferation and inflammation. Thus, visfatin derived from perivascular fat enhances proliferation in rat aortic smooth muscle cells. In endothelial cells, visfatin promotes proliferation, migration, and capillary tube formation, contributing to angiogenesis. Importantly, visfatin exerts its proliferative effects not only on the vascular wall because both adenoviral delivery and exogenous administration of visfatin increased proliferation in vitro in rat cardiomyocytes. Visfatin directly activates the ERK1/2-NF-κB axis in hVSMCs, leading to iNOS induction and a proinflammatory response. In addition, visfatin acts on endothelial cells and stimulates NF-κB−mediated induction of the adhesion molecules VCAM-1, ICAM-1, and E-selectin, finally upregulating IL-6, IL-8, and MCP-1 secretion and CCR2 expression as well as NADPH oxidase activation. Moreover, visfatin exerts immunomodulatory actions on monocytes by enhancing both TNF-α and IL-8 secretion, inhibiting apoptosis in neutrophils, and promoting macrophage survival. Visfatin contributes to extracellular matrix remodeling in the vascular wall and most likely promotes cardiac and renal fibrosis through the induction of classical profibrotic mediators such as TGF-β or PAI-1 or by directly upregulating procollagen and MMP secretion (Romacho et al., 2013).

Although initially described as a promising physiological insulin-mimetic derived from the visceral fat depot, subsequent in vitro clinical studies now point toward a role for visfatin as a biomarker or even a predictor of CVDs related to metabolic diseases, a setting with dysfunctional adipose tissue. Importantly, in vitro and ex vivo studies have demonstrated that visfatin exerts direct deleterious actions on the CV system. Thus, visfatin might play a role in CV complications related to metabolic dysfunction. Future studies directed at this topic will show if visfatin is a potential novel drug target for prevention of CVD.

OMENTIN AND APELIN

Omentin was discovered as a protein expressed by Paneth cells in the intestine. It is found within carbohydrate complexes of bacterial walls and plays a role in the defense against pathogenic bacteria. Omentin is known as an adipokine since 2005 when it was identified by a cDNA library screening approach of omental adipose tissue (Tan et al., 2010).

As indicated by its name, omentin is more abundantly expressed in visceral adipose tissue as compared with subcutaneous fat. Omentin is also predominantly expressed in the stroma—vascular fraction of adipose tissue, and data show that the expression of omentin is significantly depressed in adipose tissue from obese patients. Accordingly, serum omentin is decreased in obesity and increased in patients after weight loss. This property of omentin is similar to that of adiponectin, one of the few adipokines that is decreased in the obese state. Similar to adiponectin, omentin exerts an antiinflammatory function with beneficial effects in the CV system.

One of the first adipokine functions of omentin was the enhancement of insulin-stimulated glucose uptake in adipocytes in an autocrine/paracrine way. Subsequent studies showed that omentin acts as a vasoactive and cardioprotective factor (Yamawaki, 2011). Thus, on isolated vessels omentin inhibits noradrenaline-induced contraction. Omentin also induces an endothelium-dependent vasorelaxation in both aorta and mesenteric arteries that is mediated by NO synthesis via phosphorylation of eNOS. Based on these data, a recent study described omentin as a novel marker for endothelial dysfunction in vivo that correlates with endothelium-dependent vasodilation even after adjustment for CRP (Moreno-Navarrete et al., 2011). Circulating omentin levels were also found to be lower in type 2 diabetic patients and positively correlating with left ventricular diastolic function (Greulich et al., 2013). Interestingly, omentin was also detected in epicardial fat and is highly expressed in this tissue when compared with pericardial and subcutaneous fat. In epicardial adipose tissue of type 2 diabetic patients, omentin expression is substantially reduced. Omentin exerts cardioprotective effects on cardiomyocytes, as it prevents contractile dysfunction and improves cardiomyocyte insulin signaling (Greulich et al., 2013). The reduction in omentin expression in epicardial adipose tissue may play a role in the development of cardiac dysfunction in type 2 diabetic patients because epicardial fat directly interacts with adjacent cardiomyocytes (see Chapter 2). It is important to note that antidiabetic drugs such as metformin and TZDs significantly increase omentin serum concentrations, potentially contributing to their antidiabetic action and beneficial effects on the CV system. As for metformin, it has been shown that increased omentin levels after the treatment might contribute to lower inflammatory and angiogenic potential of patient serum on endothelial cells.

The adipocytokine apelin, like omentin and adiponectin, was found to exert cardioprotective effects in addition to its effects on glucose and lipid metabolism (Barnes et al., 2010). In addition to adipose tissue, apelin is expressed in many other tissues including the endothelial cells, the myocardium, and the brain. Apelin was first described in 1998, and it triggers downstream signaling by its binding to the G-protein−coupled APJ receptor (Huang et al., 2017; Alipour et al., 2017; Flahault et al., 2017; Lv et al., 2017; Wu et al., 2017a,b; Lu et al., 2017; Chen et al., 2016; Salska and Chizynski, 2016; Hu et al., 2016; Yang et al., 2016; Zhou et al., 2016; Huang et al., 2016). Apelin is considered as an adipokine with direct beneficial effects on both CV cells and cardiomyocytes. Thus, apelin promotes endothelium-dependent nitric oxide−mediated vasodilation, protects against atheroma, counteracts myocardial infarction, and represents one of the most potent endogenous positive inotropic agents. This view is strongly supported by observations in apelin-KO mice (Kuba et al., 2007). These animals exhibit augmented cardiac dysfunction and altered cardiac remodeling under different conditions. There is great interest in the apelin−APJ pathway because it may represent a novel target for prevention of heart failure, myocardial ischemia, and vascular disease.

FATTY LIVER AND ORGAN CROSSTALK

In the previous chapters, major emphasis was on adipose tissue and muscle regarding their secretory function and the complex crosstalk scenario involving adipokines and myokines. Compared with these two tissues, the liver has gained much less attention although it is now generally accepted that so-called hepatokines also play an important role in the regulation of metabolic homeostasis (Jung et al., 2016; Lebensztejn et al., 2016; Meex and Watt, 2017; Oh et al., 2016; Panera et al., 2016; Stefan and Haring, 2013; Yoo and Choi, 2015). Based on the available data it appears, however, that specific hepatokines are key players in states of liver dysfunction such as NAFLD playing a less important role under physiological conditions. A detailed description of hepatokines is beyond the scope of this chapter that primarily addresses the crosstalk pathways underlying NAFLD, a major chronic liver disorder with a huge prevalence and substantial impact for the development of related chronic diseases such as T2D, CVD, and chronic kidney disease.

NAFLD refers to an excessive accumulation of fat in the liver and must be considered as a spectrum of progressive liver diseases, including simple steatosis (accumulation of fat in a cell), nonalcoholic steatohepatitis, progression to fibrosis, and ultimately cirrhosis, in the absence of excessive alcohol consumption (Anstee and Day, 2013; Anstee et al., 2013; Dongiovanni et al., 2013). NAFLD is considered as the hepatic manifestation of the metabolic syndrome, and it is currently considered as a multisystem disease, also affecting extrahepatic organs and homeostatic pathways (Byrne and Targher, 2015). NAFLD is a paradigm of ectopic fat accumulation accompanied by low-grade chronic inflammation, and it must now be considered as one of the most common public health problems with a global dimension. The prevalence of NAFLD has been continuously rising, being 30%–40% in men and 15%–20% in women (Vernon et al., 2011). In patients with T2D, the prevalence of NAFLD may be even as high as 70% (Blachier et al., 2013). This reflects the strong association of NAFLD with obesity and insulin resistance, major risk factors for the development of T2D. Accumulation of lipids decreases the efficacy of insulin signaling, induces ER stress, and finally leads to activation of JNKs and NF-κB, two serine kinases and key regulators of inflammatory pathways. Specifically, the serine phosphorylation of elements of the insulin signaling machinery reduces the signal flux to the downstream effector systems (Nandipati et al., 2017; Rehman and Akash, 2016). There is now solid evidence from epidemiological, clinical, and basic studies that NAFLD is both a marker and a causal link to major extrahepatic chronic complications such as T2D, CVD, and chronic kidney disease (Byrne and Targher, 2015). Importantly, a continuous interorgan crosstalk underlies the processes involved in NAFLD pathogenesis, and different organs such as the gut, hypothalamus, and adipose tissue are involved in this complex process. A schematic presentation of this crosstalk setting was recently summarized by Zhang et al. (2017) and is depicted in Fig. 5.1.

As pointed out before, NAFLD is strongly associated to the metabolic syndrome, specifically to obesity and insulin resistance. Principally, hepatic steatosis reflects an imbalance between triacylglycerides (TAGs) influx and efflux. Fatty acids required for TAG formation are derived from adipose tissue lipolysis, de novo lipogenesis, and diet. In addition to the dysregulation of several proteins associated with lipid droplets and the ER, the hepatic overload with TAGs is known to result in an increase in free fatty acids, cholesterol, and other lipid metabolites, collectively called lipotoxicity (Marra and Svegliati-Baroni, 2017; Mota et al., 2016). This process is coupled to mitochondrial dysfunction, oxidative stress, and the formation of reactive oxygen species (ROS) as well as ER stress–related mechanisms (Buzzetti et al., 2016). Mitochondrial dysfunction is thought to play a key role in the pathophysiology of NAFLD because of the generation of lipid-derived toxic metabolites and the overproduction of ROS (Mello et al., 2016).

Obesity and adipose tissue dysfunction play a key role in the development of NAFLD owing to (1) augmented delivery of free fatty acids to hepatocytes and increased formation of hepatic TAGs and (2) the overproduction of proinflammatory adipokines such as resistin, IL-6, TNFα, and many others (Zhang et al., 2017). Thus,

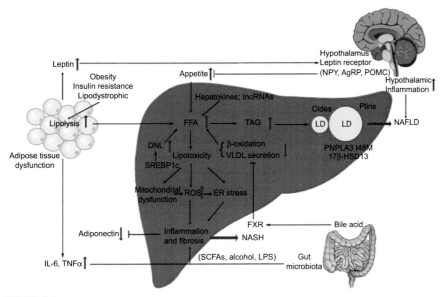

FIGURE 5.1

The "crosstalk" between liver and peripheral organs in the pathogenesis of nonalcoholic fatty liver disease (NAFLD). The impairment of the hypothalamic signaling pathway due to mutations (*leptin receptor and MC4R*) by affecting the appetite or inflammation leads to the development of obesity and NAFLD. Dysfunction of adipose tissue in obesity, lipodystrophy, or insulin resistance provides a source of excess fat and release of adipokines such as *leptin, adiponectin*, and *resistin* and proinflammatory cytokines such as tumor necrosis factor α (TNFα) and interleukin (IL)-6 that participated in the pathogenesis of NAFLD. In addition, emerging evidence suggests that an altered gut permeability consequently affects circulating levels of molecules such as lipopolysaccharide (LPS), FFA, and bile acid and the release of proinflammatory cytokines by the regulation of toll-like receptor (TLR) and FXR, further influencing the development and progression of NAFLD, recognized as effect of gut—liver axis. In the liver, the dysregulation of lipid de novo lipogenesis and imbalance of lipid influx and efflux cause lipotoxicity and may result in mitochondrial dysfunction, overproduction of reactive oxygen species (ROS), and endoplasmic reticulum (ER) stress as well as the consequent activation of inflammatory responses, thus influencing the risk of progression of NAFLD to NASH, as observed in obesity and insulin resistance. *NPY*, neuropeptide Y; *lncRNA*, long noncoding RNA; *TAG*, triacylglyceride.

Reproduced from Zhang, X, Ji, X, Wang, Q, Li, J.Z., February 2018. New insight into inter-organ crosstalk contributing to the pathogenesis of non-alcoholic fatty liver disease (NAFLD). Protein Cell 9 (2), 164–177. https://doi.org/10.1007/s13238-017-0436-0 (Epub 2017 Jun 22).

TNFα promotes the development of NAFLD and is a predictor of NASH correlating with advanced stages (Diehl, 2004). Clinical studies showed a tight association of inflammation and fibrosis with augmented levels of circulating IL-6 (van der Poorten et al., 2008). A major role in the pathogenesis of NAFLD can be allocated to adiponectin. As described in detail in Chapter 2, this beneficial adipokine is reduced in obese and type 2 diabetic subjects. It has been shown that adiponectin improves hepatic insulin resistance and exerts antiinflammatory and hepatoprotective functions (Kadowaki et al., 2006). It is conceivable that reduced levels of adiponectin, as observed in obesity, lead to hepatic steatosis, inflammation, and fibrogenesis (Tsochatzis et al., 2006).

In addition to T2D, CVD is a major extrahepatic complication tightly associated with NAFLD (Lonardo et al., 2016). Most likely, adipose tissue dysfunction plays an important role in linking these two disorders. One potential mechanism by which NAFLD increases CVD risk would be based on the development of inflamed visceral fat. It has been suggested that NAFLD can be considered as a sensitive marker of adipose tissue dysfunction that is more relevant to CV outcome than simply adipose tissue mass. An additional process relates to hepatic steatosis itself, triggering intrahepatic inflammation through activation of NF-κB pathways, finally leading to insulin resistance both locally in the liver and systemically.

It is now generally thought that NAFLD leads to liver dysfunction coupled to the release of a variety of proatherogenic, proinflammatory, and finally diabetogenic proteins such as high-sensitivity CRP (hsCRP), fibrinogen, and plasminogen activator inhibitor 1 (PAI-1), which have important roles in the development of CVD (Anstee et al., 2013). For hsCRP, known to be primarily produced by the liver, it was shown to be an independent predictor of CV events in several large studies (Lavie et al., 2009). In line with this, fibrinogen and PAI-1 also originate mostly from hepatic tissue and may augment atherothrombosis. It is therefore conceivable that enhanced liver-secreted factors in NAFLD play an important role in the pathogenesis of systemic inflammation and atherosclerosis. Therefore, the liver is definitely involved and plays an important role in the induction of systemic inflammation. On top, the liver is a key target organ of various inflammatory mediators released from dysfunctional adipose tissue. Upcoming evidence now suggests that a group of predominantly liver-derived proteins called hepatokines may directly affect glucose and lipid metabolism, as described in detail for adipokines and myokines in Chapter 2 and 3. Currently, fetuin-A, fibroblast growth factor 21 (FGF-21), and selenoprotein P (SeP) gained considerable interest as hepatokines. In the subsequent subchapters we will briefly address these hepatokines and emphasize their specific role in the development of CVD. For a recent review on this topic, see Jung et al. (2016).

FIBROBLAST GROWTH FACTOR 21

FGF-21 is a 181-amino-acid peptide hormone that is expressed in several organs but primarily secreted by the liver and acts as a potent metabolic regulator (Dushay et al., 2010). Expression of the human FGF-21 gene is regulated by PPARα during starvation, whereas FGF-21 controls PPARγ activity after feeding (Galman et al., 2008; Dutchak et al., 2012). FGF-21—KO mice display defects in PPARγ signaling including decreased body fat and attenuation of PPARγ-dependent gene expression. Moreover, FGF-21−KO mice are refractory to both the beneficial insulin-sensitizing effects and the detrimental weight gain and edema side effects of the PPARγ agonist rosiglitazone (Dutchak et al., 2012). Circulating FFAs, a characteristic feature of fasting, and several stress conditions such as hepatic injury and/or disease stimulate secretion of FGF-21 into the circulation (Yu et al., 2011). Application of FGF-21 to animals and humans has been shown to decrease body weight and the level of circulating triglycerides and low-density lipoprotein (LDL) cholesterol, and further to improve insulin sensitivity (Jeon et al., 2016). It was shown that FGF-21 treatment induces basal and insulin-stimulated glucose uptake in human skeletal muscle cells through upregulation of glucose transporter-1 (Mashili et al., 2011). Data from a clinical study suggest that treatment with LY2405319, an analogue of FGF-21, results in significant improvements in dyslipidemia of obese human subjects with T2D (Gaich et al., 2013). Thus, treatment with this compound produced significant improvements in dyslipidemia, including decreases in LDL cholesterol and triglycerides and increases in high-density lipoprotein cholesterol and a shift to a potentially less atherogenic apolipoprotein concentration profile. Favorable effects on body weight, fasting insulin, and adiponectin were also detected. However, only a trend toward glucose lowering was observed. It was concluded that FGF-21 is bioactive in humans and that FGF-21−based therapies may be effective for the treatment of selected metabolic disorders (Gaich et al., 2013). FGF-21 stimulates lipolysis in *db/db* mice and obese humans, and individuals with markers of the metabolic syndrome have increased serum FGF-21 levels to compensate for the abnormal metabolic status (Dostalova et al., 2009).

It is now generally thought that FGF-21 plays an important role in CVD. FGF-21−KO mice have increased cardiac mass and impaired cardiac function, which was improved by treatment with FGF-21, suggesting its protective role against hypertrophic insults (Planavila et al., 2013). Studies conducted with cultured cardiac microvascular endothelial cells showed an upregulation of FGF-21 when the cells were incubated with oxidized LDL, indicating that FGF-21 might be secreted by endothelial cells in response to stress. It was also suggested that elevated levels of FGF-21 may be a signal of endothelial cell injury. In accordance with these animal studies, observations in humans suggested that elevated serum FGF-21 levels are associated with carotid atherosclerosis independent of established CVD risk factors (Chow et al., 2013). It is therefore reasonable to assume that FGF-21 is an attractive target for the diagnosis and treatment of obesity and related diseases, including CVD.

FETUIN-A

Fetuin-A is a 64 kDa glycoprotein that is preferentially expressed and released by the liver (Roshanzamir et al., 2017; Vashist et al., 2017). Fetuin-A is a natural inhibitor of the insulin receptor tyrosine kinase; it is therefore tightly associated with insulin resistance (Dasgupta et al., 2010). In addition to its direct effects on the insulin receptor, fetuin-A promotes insulin resistance by triggering a proinflammatory state. Fetuin-A treatment increases expression of proinflammatory cytokines, and at the same time it reduces adiponectin expression in both adipocytes and monocytes (Hennige et al., 2008). These authors reported that fetuin-A treatment induced TNFα and IL-1β mRNA expression in THP1 cells ($P < .05$). Treatment of mice with fetuin-A, analogously, resulted in a marked increase in adipose tissue TNFα mRNA as well as IL6 expression (27- and 174-fold, respectively). These effects were accompanied by a decrease in adipose tissue adiponectin mRNA expression and lower circulating adiponectin levels ($P < .05$, both). Furthermore, fetuin-A repressed ADIPOQ mRNA expression of human in vitro differentiated adipocytes ($P < .02$) and induced inflammatory cytokine expression. In plasma of humans, fetuin-A correlated positively with hsCRP, a marker of subclinical inflammation (r = 0.26, $P = .01$), and negatively with total (r = -0.28, $P = .02$) and, particularly, high-molecular-weight adiponectin (r = -0.36, $P = .01$). It was concluded that fetuin-A induces low-grade inflammation and represses adiponectin production in animals and in humans.

Furthermore, incubation of HepG2 cells or rat hepatocytes with palmitate stimulates binding of NF-κB to the fetuin-A promoter, thereby augmenting fetuin-A synthesis and secretion (Dasgupta et al., 2010). Based on these findings it is thought that fetuin-A might directly cause insulin resistance and modulate inflammatory reactions, leading to various metabolic disturbances. In agreement, many epidemiologic studies have observed elevated levels of circulating fetuin-A in obesity and related metabolic diseases including T2D, metabolic syndrome, and NAFLD (Weikert et al., 2008; Stefan et al., 2008a,b).

However, the relationship between serum fetuin-A and CVD risk is more complicated, and additional processes need to be considered. Fetuin-A can bind Ca^{2+}, which results in reduced ectopic calcification (Mori et al., 2012). In studies examining patients with chronic kidney diseases, the level of fetuin-A was found to be inversely associated with calcification scores, CV events, and CV mortality (Evrard et al., 2015). In line with this, nondiabetics with a higher fetuin-A level have decreased risks of incident CVD and CVD-related mortality. However, patients with T2D with higher fetuin-A levels have increased risks of incident CVD and CVD-related mortality (Jensen et al., 2013). A potential pathway by which fetuin-A can promote atherosclerosis in CVD patients is induction of insulin resistance and stimulation of cytokine expression in monocytes that participate in the inflammation. Future studies need to address more specifically the contribution of fetuin-A to the development of CVD in patients with insulin resistance and T2D.

SELENOPROTEIN P

SeP is a 42 kDa glycoprotein that is mainly produced by the liver and secreted into the circulation (Burk and Hill, 2005). SeP gained considerable interest when it was identified as a hepatokine associated with insulin resistance in humans (Misu et al., 2010). These authors used serial analysis of gene expression (SAGE) and DNA chip methods and found that hepatic SeP mRNA levels correlated with insulin resistance in humans. Administration of purified SeP impaired insulin signaling and dysregulated glucose metabolism in both hepatocytes and myocytes. Conversely, both genetic deletion and RNA interference—mediated knockdown of SeP improved systemic insulin sensitivity and glucose tolerance in mice. The metabolic actions of SeP were mediated, at least partly, by inactivation of adenosine monophosphate—activated protein kinase (AMPK). It was concluded that SeP plays a role in the regulation of glucose metabolism and insulin sensitivity and that SeP may be a therapeutic target for T2D. It was also reported that patients with T2D and those with NAFLD have higher serum SeP levels than healthy controls (Yang et al., 2011). So far, there have been very few studies on the relationship of SeP with CVD. Yang et al. (2011) were the first to show that circulating SeP levels have an independent association with carotid intima—media thickness even after adjustment for other confounding factors. In an interesting paper, Ishikura et al. (2014) showed that physiological concentrations of SeP inhibit VEGF-stimulated cell proliferation, tubule formation, and migration in human umbilical vein endothelial cells. This resulted in an impaired angiogenesis and delay of wound closure in mice overexpressing SeP. The precise role of SeP in the development of CVD is an interesting and important topic of ongoing research, and SeP may turn out as a new drug target and/or biomarker of CVD.

SEDENTARY LIFESTYLE AND METABOLIC DISEASES

As already described in Chapter 2, enlargement of adipose tissue mass as a consequence of sedentary lifestyle and excessive energy intake results in a local inflammatory response in visceral adipose tissue, which is paralleled by infiltration of immune cells, and local and systemic increases of proinflammatory cytokines and adipokines. Thus, obesity, insulin resistance, and T2D are associated with chronic low-grade systemic inflammation, and regular moderate physical activity protects against a number of chronic diseases that are characterized by inflammation (Ellingsgaard et al., 2011; Ueda et al., 2009a,b). While regular physical exercise has been shown to reduce basal levels of inflammatory markers, studies have also shown that physical inactivity and sedentary lifestyle result in elevated levels of inflammatory markers (Henningsen et al., 2010; Shah et al., 2011a). Thus, in addition to changes of the adipose tissue secretome, an altered muscle secretome may play an important role in the pathophysiology of several metabolic diseases (Carson, 2017; Kostrominova, 2016; Oh et al., 2016; Pillon et al., 2013; So et al., 2014; Weigert et al., 2014). We provide here a brief summary of the antiinflammatory effects of exercise and focus on the specific role of certain myokines in the context of obesity and T2D.

THE ANTIINFLAMMATORY EFFECTS OF EXERCISE

It has been observed that regular physical activity is of key importance to reduce inflammation and to improve insulin resistance. Cross-sectional studies have shown a strong inverse association between the level of physical activity and systemic low-grade inflammation (Lee et al., 2013; Sell et al., 2012; Gleeson, 2007). These observations might be explained by an antiinflammatory effect of regular exercise, which could be mediated via different pathways. First, regular exercise programs and an active lifestyle result in enhanced energy usage, thus leading to reduction of body weight and visceral fat mass, which is an important source of proinflammatory cytokines (Pedersen and Saltin, 2006). This may explain, at least partly, the reduction of inflammatory markers such as CRP after long-term exercise (Mendham et al., 2011; Donges et al., 2010). In addition, it has been shown that exercise decreases expression of TLR2 and TLR4 in immune cells and skeletal muscle (Christiansen et al., 2010), whereas even short-term bed rest increases expression of TLR4. This needs specific attention because activation of TLR in skeletal muscle by factors such as LPS, heat shock protein 60, or free fatty acids has been suggested to participate in the development of insulin resistance (Liang et al., 2013a,b). It is also possible that physical activity may have a beneficial effect on the function of the immune system because of exercise-induced increases in adrenaline, cortisol, and other factors with immunomodulatory effects (Drummond et al., 2013).

Starkie et al. investigated the hypothesis that acute exercise induces a direct antiinflammatory response (Starkie et al., 2003). Healthy subjects received a bolus of *Escherichia coli* LPS endotoxin to induce low-grade inflammation during resting or after 2.5 h of bicycling. In resting subjects LPS-endotoxin administration resulted in a strong increase in plasma TNFα level. Interestingly, in the exercising subjects, the TNFα response to LPS-endotoxin was totally blunted, thus supporting the idea that physical activity mediates an antiinflammatory activity. These data are further supported by in vitro experiments using electrical pulse stimulation (EPS) of human primary skeletal muscle cells, which induces contraction of the cells (Lambernd et al., 2012). Our group has demonstrated that EPS prevents inflammatory responses induced by treatment with adipocyte-conditioned medium, TNFα, or MCP-1 owing to blocking of the proinflammatory NF-κB signaling pathway.

One challenge for future studies will be to uncover underlying mechanisms and mediators of the beneficial effect of exercise. The investigation of myokines and their paracrine and/or endocrine effects, as detailed in the previous chapters, will provide another explanation for the antiinflammatory and immune regulatory effects of exercise.

MYOKINES AND ANTIINFLAMMATORY EFFECTS

As described before, IL-6 is acutely increased in the circulation in response to exercise but chronically elevated in serum of patients with insulin resistance, obesity, and T2D (Dandona et al., 2004; Duncan et al., 2003; Kern et al., 2001). Although skeletal muscle has been identified as the main source of exercise-related IL-6 release, the expanding visceral adipose tissue is an important source of circulating IL-6 in conditions of obesity. Expression of IL-6 by macrophages within the adipose tissue is dependent on activation of the NF-κB signaling pathway, whereas intramuscular

IL-6 expression is regulated by different signaling cascades involving Ca^{2+}, nuclear factor of activated T cells, calcineurin, and p38 MAPK signaling. These observations initiated a controversial discussion about the functional role of IL-6 after exercise and in inflammatory conditions (reviewed in detail by Pedersen and Febbraio, 2008). Accumulating evidence suggests that exercise-related IL-6 released by skeletal muscle triggers an antiinflammatory cascade by inducing the production of the antiinflammatory cytokines IL-10, IL-1 receptor antagonist (IL-1RA), and soluble TNF receptor (sTNFR) (Pedersen and Febbraio, 2008). Based on existing data, it can be assumed that IL-6 inhibits the production of proinflammatory TNFα. Thus, Starkie et al. (2003) have reported that in healthy humans elevation of plasma IL-6 level by acute exercise or infusion of recombinant IL-6 blunted an LPS-endotoxin–mediated increase of TNFα. The presence of IL-10, IL-1RA, and sTNFR in the circulation after exercise contributes to the antiinflammatory effect of exercise (see Fig. 5.2). IL-10 inhibits the production of proinflammatory cytokines such as IL-1α, IL-1β, and TNFα and chemokines such as IL-8 and MIPα (Petersen and Pedersen, 2005). IL-1RA is a member of the IL-1 family, binds to the IL-1 receptor, and inhibits the intracellular signal transduction, thereby reducing the proinflammatory cascade induced by IL-1β (Dinarello, 2000).

Our group recently described chitinase-3–like protein 1 (CHI3L1 or YKL-40) as a novel myokine (Gorgens et al., 2014). Importantly, plasma CHI3L1 levels are upregulated in patients with insulin resistance and T2D independent of obesity (Nielsen et al., 2008). At present, the physiological impact and the source of circulating CHI3L1 remain largely unknown. Our in vitro experiments have shown that expression and release of CHI3L1 is stimulated by proinflammatory cytokines such as TNFα. Our data provide evidence that CHI3L1 protects myotubes from TNFα-induced insulin resistance and inflammation by inhibiting NF-κB activation, making it likely that CHI3L1 acts as an autoprotective factor that is induced on demand to protect skeletal muscle from the negative effects of TNFα (Gorgens et al., 2014). Taken together, these data indicate that myokines may be involved in mediating the antiinflammatory effects of exercise and thus in mediating the beneficial effect of exercise on health. They may play a crucial role in the protection against diseases characterized by low-grade inflammation such as insulin resistance and type 2 diabetes.

In contrast to the observation that regular moderate exercise reduces chronic inflammation, high-intensity training causes a temporary depression of various aspects of immune function and an increase in systemic inflammation for a certain postexercise period ($\sim 3–24$ h) (Gleeson, 2007). It has been shown that after a very intensive exercise such as marathon running, TNFα and IL-1β levels increase in response to muscle damage (Bernecker et al., 2013). It is well established that muscle repair and regeneration after acute muscle injury involves a tissue-remodeling and growth-promoting local inflammation. The initial inflammatory response is required for a positive outcome of muscle repair (Paulsen et al., 2012). This supports the notion that inflammation exerts multiple, both beneficial and harmful, functions in metabolic regulation.

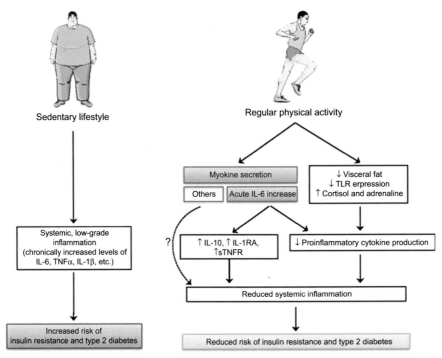

FIGURE 5.2

Impact of physical activity level on inflammatory status and risk of insulin resistance and/or type 2 diabetes (T2D). On the one hand, a sedentary lifestyle combined with obesity is often associated with systemic low-grade inflammation and an increased risk of insulin resistance and T2D. On the other hand, an active lifestyle with regular physical activity decreases the risk of developing insulin resistance and T2D. This is in part mediated by the beneficial effects of exercise on systemic inflammation status. The exercise-induced acute increase in interleukin (IL)-6 results in the enhanced release of antiinflammatory cytokines such as IL-10, IL-1 receptor antagonist (IL-1RA), and soluble tumor necrosis factor receptor (sTNFR), which reduces systemic inflammation. It is currently unknown whether other myokines may also directly or indirectly affect systemic inflammation. In addition, physical activity reduces visceral adipose tissue and toll-like receptor (TLR) expression in immune cells and skeletal muscle and increases the release of adrenaline and cortisol, which is known to have potent antiinflammatory effects. Overall these events result in the reduced production of proinflammatory cytokines, thereby contributing to reduced systemic inflammation and a decreased risk of developing insulin resistance and T2D.

In conclusion, it is obvious that regular physical activity is useful for prevention and therapy of various pathological conditions. For patients with T2D, a combination of both endurance and strength training was suggested as the most beneficial exercise training to improve glycemic control (Eckardt et al., 2014; Sigal et al., 2007). To understand the underlying mechanism of the antiinflammatory effect of exercise in more detail, future efforts may also need to focus on the endocrine effects of myokines on immune cells.

REFERENCES

Alipour, F.G., Ashoori, M.R., Pilehvar-Soltanahmadi, Y., Zarghami, N., 2017. An overview on biological functions and emerging therapeutic roles of apelin in diabetes mellitus. Diabet. Metab. Syndr. 11 (Suppl. 2), S919—S923.

Alisi, A., Carpino, G., Oliveira, F.L., Panera, N., Nobili, V., Gaudio, E., 2017. The role of tissue macrophage-mediated inflammation on NAFLD pathogenesis and its clinical implications. Mediat. Inflamm. 2017, 8162421.

Anderson, J.W., Kendall, C.W., Jenkins, D.J., 2003. Importance of weight management in type 2 diabetes: review with meta-analysis of clinical studies. J. Am. Coll. Nutr. 22, 331—339.

Anstee, Q.M., Day, C.P., 2013. The genetics of NAFLD. Nat. Rev. Gastroenterol. Hepatol. 10, 645—655.

Anstee, Q.M., Targher, G., Day, C.P., 2013. Progression of NAFLD to diabetes mellitus, cardiovascular disease or cirrhosis. Nat. Rev. Gastroenterol. Hepatol. 10, 330—344.

Argiles, J.M., Lopez-Soriano, J., Almendro, V., Busquets, S., Lopez-Soriano, F.J., 2005. Cross-talk between skeletal muscle and adipose tissue: a link with obesity? Med. Res. Rev. 25, 49—65.

Arkan, M.C., Hevener, A.L., Greten, F.R., Maeda, S., Li, Z.W., Long, J.M., Wynshaw-Boris, A., Poli, G., Olefsky, J., Karin, M., 2005. IKK-beta links inflammation to obesity-induced insulin resistance. Nat. Med. 11, 191—198.

Baldrighi, M., Mallat, Z., Li, X., 2017. NLRP3 inflammasome pathways in atherosclerosis. Atherosclerosis 267, 127—138.

Bandyopadhyay, G.K., Yu, J.G., Ofrecio, J., Olefsky, J.M., 2005. Increased p85/55/50 expression and decreased phosphatidylinositol 3-kinase activity in insulin-resistant human skeletal muscle. Diabetes 54, 2351—2359.

Barnes, G., Japp, A.G., Newby, D.E., 2010. Translational promise of the apelin—APJ system. Heart 96, 1011—1016.

Bernecker, C., Scherr, J., Schinner, S., Braun, S., Scherbaum, W.A., Halle, M., 2013. Evidence for an exercise induced increase of TNF-alpha and IL-6 in marathon runners. Scand. J. Med. Sci. Sports 23, 207—214.

Blachier, M., Leleu, H., Peck-Radosavljevic, M., Valla, D.C., Roudot-Thoraval, F., 2013. The burden of liver disease in Europe: a review of available epidemiological data. J. Hepatol. 58, 593—608.

Bosello, O., Zamboni, M., 2000. Visceral obesity and metabolic syndrome. Obes. Rev. 1, 47—56.

Brozinick Jr., J.T., Roberts, B.R., Dohm, G.L., 2003. Defective signaling through Akt-2 and -3 but not Akt-1 in insulin-resistant human skeletal muscle: potential role in insulin resistance. Diabetes 52, 935—941.

Bruce, C.R., Carey, A.L., Hawley, J.A., Febbraio, M.A., 2003. Intramuscular heat shock protein 72 and heme oxygenase-1 mRNA are reduced in patients with type 2 diabetes: evidence that insulin resistance is associated with a disturbed antioxidant defense mechanism. Diabetes 52, 2338–2345.

Burk, R.F., Hill, K.E., 2005. Selenoprotein P: an extracellular protein with unique physical characteristics and a role in selenium homeostasis. Annu. Rev. Nutr. 25, 215–235.

Busse, W.W., Calhoun, W.F., Sedgwick, J.D., 1993. Mechanism of airway inflammation in asthma. Am. Rev. Respir. Dis. 147, S20–S24.

Buzzetti, E., Pinzani, M., Tsochatzis, E.A., 2016. The multiple-hit pathogenesis of non-alcoholic fatty liver disease (NAFLD). Metabolism 65, 1038–1048.

Byrne, C.D., Targher, G., 2015. NAFLD: a multisystem disease. J. Hepatol. 62, S47–S64.

Carson, B.P., 2017. The potential role of contraction-induced myokines in the regulation of metabolic function for the prevention and treatment of type 2 diabetes. Front. Endocrinol. 8, 97.

Chen, Z., Wu, D., Li, L., Chen, L., 2016. Apelin/APJ system: a novel therapeutic target for myocardial ischemia/reperfusion injury. DNA Cell Biol. 35, 766–775.

Chow, W.S., Xu, A., Woo, Y.C., Tso, A.W., Cheung, S.C., Fong, C.H., Tse, H.F., Chau, M.T., Cheung, B.M., Lam, K.S., 2013. Serum fibroblast growth factor-21 levels are associated with carotid atherosclerosis independent of established cardiovascular risk factors. Arterioscler. Thromb. Vasc. Biol. 33, 2454–2459.

Christiansen, T., Paulsen, S.K., Bruun, J.M., Pedersen, S.B., Richelsen, B., 2010. Exercise training versus diet-induced weight-loss on metabolic risk factors and inflammatory markers in obese subjects: a 12-week randomized intervention study. Am. J. Physiol. Endocrinol. Metab. 298, E824–E831.

Cotillard, A., Poitou, C., Torcivia, A., Bouillot, J.L., Dietrich, A., Kloting, N., Gregoire, C., Lolmede, K., Bluher, M., Clement, K., 2014. Adipocyte size threshold matters: link with risk of type 2 diabetes and improved insulin resistance after gastric bypass. J. Clin. Endocrinol. Metab. 99, E1466–E1470.

Cronstein, B.N., 1992. Molecular mechanism of methotrexate action in inflammation. Inflammation 16, 411–423.

Cuthbertson, D.J., Steele, T., Wilding, J.P., Halford, J.C., Harrold, J.A., Hamer, M., Karpe, F., 2017. What have human experimental overfeeding studies taught us about adipose tissue expansion and susceptibility to obesity and metabolic complications? Int. J. Obes. 41, 853–865.

Czech, M.P., 2017. Insulin action and resistance in obesity and type 2 diabetes. Nat. Med. 23, 804–814.

Dahlman, I., Forsgren, M., Sjogren, A., Nordstrom, E.A., Kaaman, M., Naslund, E., Attersand, A., Arner, P., 2006. Downregulation of electron transport chain genes in visceral adipose tissue in type 2 diabetes independent of obesity and possibly involving tumor necrosis Factor-{alpha}. Diabetes 55, 1792–1799.

Dandona, P., Aljada, A., Bandyopadhyay, A., 2004. Inflammation: the link between insulin resistance, obesity and diabetes. Trends Immunol. 25, 4–7.

Dasgupta, S., Bhattacharya, S., Biswas, A., Majumdar, S.S., Mukhopadhyay, S., Ray, S., Bhattacharya, S., 2010. NF-kappaB mediates lipid-induced fetuin-A expression in hepatocytes that impairs adipocyte function effecting insulin resistance. Biochem. J. 429, 451–462.

Di Raimondo, D., Tuttolomondo, A., Musiari, G., Schimmenti, C., DAngelo, A., Pinto, A., 2016. Are the myokines the mediators of physical activity-induced health Benefits? Curr. Pharmaceut. Des. 22, 3622—3647.

Diehl, A.M., 2004. Tumor necrosis factor and its potential role in insulin resistance and nonalcoholic fatty liver disease. Clin. Liver Dis. 8, 619—638.

Dietze, D., Koenen, M., Rohrig, K., Horikoshi, H., Hauner, H., Eckel, J., 2002. Impairment of insulin signaling in human skeletal muscle cells by co-culture with human adipocytes. Diabetes 51, 2369—2376.

Dietze, D., Ramrath, S., Ritzeler, O., Tennagels, N., Hauner, H., Eckel, J., 2004. Inhibitor kappaB kinase is involved in the paracrine crosstalk between human fat and muscle cells. Int. J. Obes. Relat. Metab. Disord. 28, 985—992.

Dietze-Schroeder, D., Sell, H., Uhlig, M., Koenen, M., Eckel, J., 2005. Autocrine action of adiponectin on human fat cells prevents the release of insulin resistance-inducing factors. Diabetes 54, 2003—2011.

Dinarello, C.A., 2000. The role of the interleukin-1-receptor antagonist in blocking inflammation mediated by interleukin-1. N. Engl. J. Med. 343, 732—734.

Donges, C.E., Duffield, R., Drinkwater, E.J., 2010. Effects of resistance or aerobic exercise training on interleukin-6, C-reactive protein, and body composition. Med. Sci. Sports Exerc. 42, 304—313.

Dongiovanni, P., Anstee, Q.M., Valenti, L., 2013. Genetic predisposition in NAFLD and NASH: impact on severity of liver disease and response to treatment. Curr. Pharmaceut. Des. 19, 5219—5238.

Dostalova, I., Haluzikova, D., Haluzik, M., 2009. Fibroblast growth factor 21: a novel metabolic regulator with potential therapeutic properties in obesity/type 2 diabetes mellitus. Physiol. Res. 58, 1—7.

Drummond, M.J., Timmerman, K.L., Markofski, M.M., Walker, D.K., Dickinson, J.M., Jamaluddin, M., Brasier, A.R., Rasmussen, B.B., Volpi, E., 2013. Short-term bed rest increases TLR4 and IL-6 expression in skeletal muscle of older adults. Am. J. Physiol. Regul. Integr. Comp. Physiol. 305, R216—R223.

Duncan, B.B., Schmidt, M.I., Pankow, J.S., Ballantyne, C.M., Couper, D., Vigo, A., Hoogeveen, R., Folsom, A.R., Heiss, G., Atherosclerosis Risk in Communities, Study, 2003. Low-grade systemic inflammation and the development of type 2 diabetes: the atherosclerosis risk in communities study. Diabetes 52, 1799—1805.

Dushay, J., Chui, P.C., Gopalakrishnan, G.S., Varela-Rey, M., Crawley, M., Fisher, F.M., Badman, M.K., Martinez-Chantar, M.L., Maratos-Flier, E., 2010. Increased fibroblast growth factor 21 in obesity and nonalcoholic fatty liver disease. Gastroenterology 139, 456—463.

Dutchak, P.A., Katafuchi, T., Bookout, A.L., Choi, J.H., Yu, R.T., Mangelsdorf, D.J., Kliewer, S.A., 2012. Fibroblast growth factor-21 regulates PPARgamma activity and the antidiabetic actions of thiazolidinediones. Cell 148, 556—567.

Eckardt, K., Gorgens, S.W., Raschke, S., Eckel, J., 2014. Myokines in insulin resistance and type 2 diabetes. Diabetologia 57, 1087—1099.

Ellingsgaard, H., Hauselmann, I., Schuler, B., Habib, A.M., Baggio, L.L., Meier, D.T., Eppler, E., Bouzakri, K., Wueest, S., Muller, Y.D., Hansen, A.M., Reinecke, M., Konrad, D., Gassmann, M., Reimann, F., Halban, P.A., Gromada, J., Drucker, D.J., Gribble, F.M., Ehses, J.A., Donath, M.Y., 2011. Interleukin-6 enhances insulin secretion by increasing glucagon-like peptide-1 secretion from L cells and alpha cells. Nat. Med. 17, 1481—1489.

Ellulu, M.S., 2017. Obesity, cardiovascular disease, and role of vitamin C on inflammation: a review of facts and underlying mechanisms. Inflammopharmacology 25, 313−328.

Engin, A., 2017. Adipose tissue hypoxia in obesity and its impact on preadipocytes and macrophages: hypoxia hypothesis. Adv. Exp. Med. Biol. 960, 305−326.

Evrard, S., Delanaye, P., Kamel, S., Cristol, J.P., Cavalier, E., Sfbc Sn Joined Working Group on Vascular Calcifications, 2015. Vascular calcification: from pathophysiology to biomarkers. Clin. Chim. Acta 438, 401−414.

Fadini, G.P., Avogaro, A., 2011. Cardiovascular effects of DPP-4 inhibition: beyond GLP-1. Vasc. Pharmacol. 55, 10−16.

Ferrante Jr., A.W., 2007. Obesity-induced inflammation: a metabolic dialogue in the language of inflammation. J. Intern. Med. 262, 408−414.

Flahault, A., Couvineau, P., Alvear-Perez, R., Iturrioz, X., Llorens-Cortes, C., 2017. Role of the vasopressin/apelin balance and potential use of metabolically stable apelin analogs in water metabolism disorders. Front. Endocrinol. 8, 120.

Frost, R.A., Nystrom, G.J., Lang, C.H., 2004. Lipopolysaccharide stimulates nitric oxide synthase-2 expression in murine skeletal muscle and C(2)C(12) myoblasts via Toll-like receptor-4 and c-Jun NH(2)-terminal kinase pathways. Am. J. Physiol. Cell Physiol. 287, C1605−C1615.

Fruh, S.M., 2017. Obesity: risk factors, complications, and strategies for sustainable long-term weight management. J Am Assoc Nurse Pract 29, S3−S14.

Fukuhara, A., Matsuda, M., Nishizawa, M., Segawa, K., Tanaka, M., Kishimoto, K., Matsuki, Y., Murakami, M., Ichisaka, T., Murakami, H., Watanabe, E., Takagi, T., Akiyoshi, M., Ohtsubo, T., Kihara, S., Yamashita, S., Makishima, M., Funahashi, T., Yamanaka, S., Hiramatsu, R., Matsuzawa, Y., Shimomura, I., 2005. Visfatin: a protein secreted by visceral fat that mimics the effects of insulin. Science 307, 426−430.

Fukuhara, A., Matsuda, M., Nishizawa, M., Segawa, K., Tanaka, M., Kishimoto, K., Matsuki, Y., Murakami, M., Ichisaka, T., Murakami, H., Watanabe, E., Takagi, T., Akiyoshi, M., Ohtsubo, T., Kihara, S., Yamashita, S., Makishima, M., Funahashi, T., Yamanaka, S., Hiramatsu, R., Matsuzawa, Y., Shimomura, I., 2007. Retraction. Science 318, 565.

Fulop, P., Harangi, M., Seres, I., Paragh, G., 2016. Paraoxonase-1 and adipokines: potential links between obesity and atherosclerosis. Chem. Biol. Interact. 259, 388−393.

Gaich, G., Chien, J.Y., Fu, H., Glass, L.C., Deeg, M.A., Holland, W.L., Kharitonenkov, A., Bumol, T., Schilske, H.K., Moller, D.E., 2013. The effects of LY2405319, an FGF21 analog, in obese human subjects with type 2 diabetes. Cell Metabol. 18, 333−340.

Galman, C., Lundasen, T., Kharitonenkov, A., Bina, H.A., Eriksson, M., Hafstrom, I., Dahlin, M., Amark, P., Angelin, B., Rudling, M., 2008. The circulating metabolic regulator FGF21 is induced by prolonged fasting and PPARalpha activation in man. Cell Metabol. 8, 169−174.

Gao, Z., Hwang, D., Bataille, F., Lefevre, M., York, D., Quon, M.J., Ye, J., 2002. Serine phosphorylation of insulin receptor substrate 1 by inhibitor kappa B kinase complex. J. Biol. Chem. 277, 48115−48121.

Gleeson, M., 2007. Immune function in sport and exercise. J. Appl. Physiol. 103, 693−699.

Gonzalez, K., Fuentes, J., Marquez, J.L., 2017a. Physical inactivity, sedentary behavior and chronic diseases. Korean J. Fam. Med. 38, 111−115.

Gonzalez, N., Moreno-Villegas, Z., Gonzalez-Bris, A., Egido, J., Lorenzo, O., 2017b. Regulation of visceral and epicardial adipose tissue for preventing cardiovascular injuries associated to obesity and diabetes. Cardiovasc. Diabetol. 16, 44.

Goodarzi, M.O., 2017. Genetics of obesity: what genetic association studies have taught us about the biology of obesity and its complications. Lancet Diabetes Endocrinol.

Gordon, S., 2003. Alternative activation of macrophages. Nat. Rev. Immunol. 3, 23−35.

Gorgens, S.W., Eckardt, K., Elsen, M., Tennagels, N., Eckel, J., 2014. Chitinase-3-like protein 1 protects skeletal muscle from TNFalpha-induced inflammation and insulin resistance. Biochem. J. 459, 479−488.

Greulich, S., Chen, W.J., Maxhera, B., Rijzewijk, L.J., van der Meer, R.W., Jonker, J.T., Mueller, H., de Wiza, D.H., Floerke, R.R., Smiris, K., Lamb, H.J., de Roos, A., Bax, J.J., Romijn, J.A., Smit, J.W., Akhyari, P., Lichtenberg, A., Eckel, J., Diamant, M., Ouwens, D.M., 2013. Cardioprotective properties of omentin-1 in type 2 diabetes: evidence from clinical and in vitro studies. PLoS One 8, e59697.

Haasch, D., Berg, C., Clampit, J.E., Pederson, T., Frost, L., Kroeger, P., Rondinone, C.M., 2006. PKCtheta is a key player in the development of insulin resistance. Biochem. Biophys. Res. Commun. 343, 361−368.

Haslam, D.W., James, W.P., 2005. Obesity. Lancet 366, 1197−1209.

Heilig, R., Broz, P., 2017. Function and mechanism of the pyrin inflammasome. Eur. J. Immunol.

Hennige, A.M., Staiger, H., Wicke, C., Machicao, F., Fritsche, A., Haring, H.U., Stefan, N., 2008. Fetuin-A induces cytokine expression and suppresses adiponectin production. PLoS One 3, e1765.

Henningsen, J., Rigbolt, K.T., Blagoev, B., Pedersen, B.K., Kratchmarova, I., 2010. Dynamics of the skeletal muscle secretome during myoblast differentiation. Mol. Cell. Proteomics 9, 2482−2496.

Heymann, D., 2006. Autophagy: a protective mechanism in response to stress and inflammation. Curr. Opin. Invest. Drugs 7, 443−450.

Hirsch, J., Batchelor, B., 1976. Adipose tissue cellularity in human obesity. Clin. Endocrinol. Metabol. 5, 299−311.

Hu, H., He, L., Li, L., Chen, L., 2016. Apelin/APJ system as a therapeutic target in diabetes and its complications. Mol. Genet. Metabol. 119, 20−27.

Hu, Y., Liu, H., Simpson, R.W., Dear, A.E., 2013. GLP-1-dependent and independent effects and molecular mechanisms of a dipeptidyl peptidase 4 inhibitor in vascular endothelial cells. Mol. Biol. Rep. 40, 2273−2279.

Huang, S., Chen, L., Lu, L., Li, L., 2016. The apelin-APJ axis: a novel potential therapeutic target for organ fibrosis. Clin. Chim. Acta 456, 81−88.

Huang, Z., Wu, L., Chen, L., 2017. Apelin/APJ system: a novel potential therapy target for kidney disease. J. Cell. Physiol.

IDF, 2003. Diabetes Atlas. International Diabetes Federation, Brussels.

Ishikura, K., Misu, H., Kumazaki, M., Takayama, H., Matsuzawa-Nagata, N., Tajima, N., Chikamoto, K., Lan, F., Ando, H., Ota, T., Sakurai, M., Takeshita, Y., Kato, K., Fujimura, A., Miyamoto, K., Saito, Y., Kameo, S., Okamoto, Y., Takuwa, Y., Takahashi, K., Kidoya, H., Takakura, N., Kaneko, S., Takamura, T., 2014. Selenoprotein P as a diabetes-associated hepatokine that impairs angiogenesis by inducing VEGF resistance in vascular endothelial cells. Diabetologia 57, 1968−1976.

Jager, J., Gremeaux, T., Cormont, M., Le Marchand-Brustel, Y., Tanti, J.F., 2007. Interleukin-1beta-induced insulin resistance in adipocytes through down-regulation of insulin receptor substrate-1 expression. Endocrinology 148, 241−251.

Jang, Y., Lee, J.H., Wang, Y., Sweeney, G., 2012. Emerging clinical and experimental evidence for the role of lipocalin-2 in metabolic syndrome. Clin. Exp. Pharmacol. Physiol. 39, 194–199.

Jensen, M.K., Bartz, T.M., Mukamal, K.J., Djousse, L., Kizer, J.R., Tracy, R.P., Zieman, S.J., Rimm, E.B., Siscovick, D.S., Shlipak, M., Ix, J.H., 2013. Fetuin-A, type 2 diabetes, and risk of cardiovascular disease in older adults: the cardiovascular health study. Diabetes Care 36, 1222–1228.

Jeon, J.Y., Choi, S.E., Ha, E.S., Kim, T.H., Jung, J.G., Han, S.J., Kim, H.J., Kim, D.J., Kang, Y., Lee, K.W., 2016. Association between insulin resistance and impairment of FGF21 signal transduction in skeletal muscles. Endocrine 53, 97–106.

Jung, T.W., Yoo, H.J., Choi, K.M., 2016. Implication of hepatokines in metabolic disorders and cardiovascular diseases. BBA Clin 5, 108–113.

Kachur, S., Lavie, C.J., de Schutter, A., Milani, R.V., Ventura, H.O., 2017. Obesity and cardiovascular diseases. Minerva Med. 108, 212–228.

Kadowaki, T., Yamauchi, T., Kubota, N., Hara, K., Ueki, K., Tobe, K., 2006. Adiponectin and adiponectin receptors in insulin resistance, diabetes, and the metabolic syndrome. J. Clin. Invest. 116, 1784–1792.

Karalis, K.P., Giannogonas, P., Kodela, E., Koutmani, Y., Zoumakis, M., Teli, T., 2009. Mechanisms of obesity and related pathology: linking immune responses to metabolic stress. FEBS J. 276, 5747–5754.

Kelley, D.E., He, J., Menshikova, E.V., Ritov, V.B., 2002. Dysfunction of mitochondria in human skeletal muscle in type 2 diabetes. Diabetes 51, 2944–2950.

Kern, P.A., Ranganathan, S., Li, C., Wood, L., Ranganathan, G., 2001. Adipose tissue tumor necrosis factor and interleukin-6 expression in human obesity and insulin resistance. Am. J. Physiol. Endocrinol. Metab. 280, E745–E751.

Kim, J.K., Jin, H.S., Suh, H.W., Jo, E.K., 2017. Negative regulators and their mechanisms in NLRP3 inflammasome activation and signaling. Immunol. Cell Biol. 95, 584–592.

Kopitar-Jerala, N., 2017. The role of interferons in inflammation and inflammasome activation. Front. Immunol. 8, 873.

Kostrominova, T.Y., 2016. Role of myokines in the maintenance of whole-body metabolic homeostasis. Minerva Endocrinol. 41, 403–420.

Kowalski, G.M., Nicholls, H.T., Risis, S., Watson, N.K., Kanellakis, P., Bruce, C.R., Bobik, A., Lancaster, G.I., Febbraio, M.A., 2011. Deficiency of haematopoietic-cell-derived IL-10 does not exacerbate high-fat-diet-induced inflammation or insulin resistance in mice. Diabetologia 54, 888–899.

Krebs, M., Roden, M., 2005. Molecular mechanisms of lipid-induced insulin resistance in muscle, liver and vasculature. Diabetes Obes. Metabol. 7, 621–632.

Kuba, K., Zhang, L., Imai, Y., Arab, S., Chen, M., Maekawa, Y., Leschnik, M., Leibbrandt, A., Markovic, M., Schwaighofer, J., Beetz, N., Musialek, R., Neely, G.G., Komnenovic, V., Kolm, U., Metzler, B., Ricci, R., Hara, H., Meixner, A., Nghiem, M., Chen, X., Dawood, F., Wong, K.M., Sarao, R., Cukerman, E., Kimura, A., Hein, L., Thalhammer, J., Liu, P.P., Penninger, J.M., 2007. Impaired heart contractility in Apelin gene-deficient mice associated with aging and pressure overload. Circ. Res. 101, e32–42.

Lambernd, S., Taube, A., Schober, A., Platzbecker, B., Gorgens, S.W., Schlich, R., Jeruschke, K., Weiss, J., Eckardt, K., Eckel, J., 2012. Contractile activity of human skeletal muscle cells prevents insulin resistance by inhibiting pro-inflammatory signalling pathways. Diabetologia 55, 1128–1139.

Lamers, D., Famulla, S., Wronkowitz, N., Hartwig, S., Lehr, S., Ouwens, D.M., Eckardt, K., Kaufman, J.M., Ryden, M., Muller, S., Hanisch, F.G., Ruige, J., Arner, P., Sell, H., Eckel, J., 2011. Dipeptidyl peptidase 4 is a novel adipokine potentially linking obesity to the metabolic syndrome. Diabetes 60, 1917−1925.

Lappas, M., Yee, K., Permezel, M., Rice, G.E., 2005. Sulfasalazine and BAY 11-7082 interfere with the nuclear factor-kappa B and I kappa B kinase pathway to regulate the release of proinflammatory cytokines from human adipose tissue and skeletal muscle in vitro. Endocrinology 146, 1491−1497.

Larsen, C.M., Faulenbach, M., Vaag, A., Volund, A., Ehses, J.A., Seifert, B., Mandrup-Poulsen, T., Donath, M.Y., 2007. Interleukin-1-receptor antagonist in type 2 diabetes mellitus. N. Engl. J. Med. 356, 1517−1526.

Lavie, C.J., Milani, R.V., Verma, A., O'Keefe, J.H., 2009. C-reactive protein and cardiovascular diseases−is it ready for primetime? Am. J. Med. Sci. 338, 486−492.

Lebensztejn, D.M., Flisiak-Jackiewicz, M., Bialokoz-Kalinowska, I., Bobrus-Chociej, A., Kowalska, I., 2016. Hepatokines and non-alcoholic fatty liver disease. Acta Biochim. Pol. 63, 459−467.

Lee, Y.S., Morinaga, H., Kim, J.J., Lagakos, W., Taylor, S., Keshwani, M., Perkins, G., Dong, H., Kayali, A.G., Sweet, I.R., Olefsky, J., 2013. The fractalkine/CX3CR1 system regulates beta cell function and insulin secretion. Cell 153, 413−425.

Li, F., Li, Y., Duan, Y., Hu, C.A., Tang, Y., Yin, Y., 2017. Myokines and adipokines: involvement in the crosstalk between skeletal muscle and adipose tissue. Cytokine Growth Factor Rev. 33, 73−82.

Liang, H., Hussey, S.E., Sanchez-Avila, A., Tantiwong, P., Musi, N., 2013a. Effect of lipopolysaccharide on inflammation and insulin action in human muscle. PLoS One 8, e63983.

Liang, H., Tantiwong, P., Sriwijitkamol, A., Shanmugasundaram, K., Mohan, S., Espinoza, S., Defronzo, R.A., Dube, J.J., Musi, N., 2013b. Effect of a sustained reduction in plasma free fatty acid concentration on insulin signalling and inflammation in skeletal muscle from human subjects. J. Physiol. 591, 2897−2909.

Liu, J.T., Song, E., Xu, A., Berger, T., Mak, T.W., Tse, H.F., Law, I.K., Huang, B., Liang, Y., Vanhoutte, P.M., Wang, Y., 2012. Lipocalin-2 deficiency prevents endothelial dysfunction associated with dietary obesity: role of cytochrome P450 2C inhibition. Br. J. Pharmacol. 165, 520−531.

Liu, S.F., Malik, A.B., 2006. NF-kappa B activation as a pathological mechanism of septic shock and inflammation. Am. J. Physiol. Lung Cell Mol. Physiol. 290, L622−L645.

Lofgren, I., Herron, K., Zern, T., West, K., Patalay, M., Shachter, N.S., Koo, S.I., Fernandez, M.L., 2004. Waist circumference is a better predictor than body mass index of coronary heart disease risk in overweight premenopausal women. J. Nutr. 134, 1071−1076.

Lonardo, A., Ballestri, S., Guaraldi, G., Nascimbeni, F., Romagnoli, D., Zona, S., Targher, G., 2016. Fatty liver is associated with an increased risk of diabetes and cardiovascular disease - evidence from three different disease models: NAFLD, HCV and HIV. World J. Gastroenterol. 22, 9674−9693.

Lu, L., Wu, D., Li, L., Chen, L., 2017. Apelin/APJ system: a bifunctional target for cardiac hypertrophy. Int. J. Cardiol. 230, 164−170.

Lv, X., Kong, J., Chen, W.D., Wang, Y.D., 2017. The role of the apelin/APJ system in the regulation of liver disease. Front. Pharmacol. 8, 221.

Marra, F., Svegliati-Baroni, G., 2017. Lipotoxicity and the gut-liver axis in NASH pathogenesis. J. Hepatol.

Mashili, F.L., Austin, R.L., Deshmukh, A.S., Fritz, T., Caidahl, K., Bergdahl, K., Zierath, J.R., Chibalin, A.V., Moller, D.E., Kharitonenkov, A., Krook, A., 2011. Direct effects of FGF21 on glucose uptake in human skeletal muscle: implications for type 2 diabetes and obesity. Diabetes Metab Res Rev 27, 286–297.

Matsuzawa, Y., Funahashi, T., Nakamura, T., 2011. The concept of metabolic syndrome: contribution of visceral fat accumulation and its molecular mechanism. J. Atherosclerosis Thromb. 18, 629–639.

Matsuzawa, Y., Nakamura, T., Shimomura, I., Kotani, K., 1995. Visceral fat accumulation and cardiovascular disease. Obes. Res. 3 (Suppl. 5), 645S–647S.

Matsuzawa, Y., Shimomura, I., Nakamura, T., Keno, Y., Tokunaga, K., 1994. Pathophysiology and pathogenesis of visceral fat obesity. Diabetes Res. Clin. Pract. 24 (Suppl.), S111–S116.

Meex, R.C.R., Watt, M.J., 2017. Hepatokines: linking nonalcoholic fatty liver disease and insulin resistance. Nat. Rev. Endocrinol. 13, 509–520.

Mello, T., Materozzi, M., Galli, A., 2016. PPARs and mitochondrial metabolism: from NAFLD to HCC. PPAR Res. 2016, 7403230.

Mendham, A.E., Donges, C.E., Liberts, E.A., Duffield, R., 2011. Effects of mode and intensity on the acute exercise-induced IL-6 and CRP responses in a sedentary, overweight population. Eur. J. Appl. Physiol. 111, 1035–1045.

Misu, H., Takamura, T., Takayama, H., Hayashi, H., Matsuzawa-Nagata, N., Kurita, S., Ishikura, K., Ando, H., Takeshita, Y., Ota, T., Sakurai, M., Yamashita, T., Mizukoshi, E., Yamashita, T., Honda, M., Miyamoto, K., Kubota, T., Kubota, N., Kadowaki, T., Kim, H.J., Lee, I.K., Minokoshi, Y., Saito, Y., Takahashi, K., Yamada, Y., Takakura, N., Kaneko, S., 2010. A liver-derived secretory protein, selenoprotein P, causes insulin resistance. Cell Metabol. 12, 483–495.

Mootha, V.K., Lindgren, C.M., Eriksson, K.F., Subramanian, A., Sihag, S., Lehar, J., Puigserver, P., Carlsson, E., Ridderstrale, M., Laurila, E., Houstis, N., Daly, M.J., Patterson, N., Mesirov, J.P., Golub, T.R., Tamayo, P., Spiegelman, B., Lander, E.S., Hirschhorn, J.N., Altshuler, D., Groop, L.C., 2003. PGC-1alpha-responsive genes involved in oxidative phosphorylation are coordinately downregulated in human diabetes. Nat. Genet. 34, 267–273.

Moreno-Navarrete, J.M., Ortega, F., Castro, A., Sabater, M., Ricart, W., Fernandez-Real, J.M., 2011. Circulating omentin as a novel biomarker of endothelial dysfunction. Obesity 19, 1552–1559.

Mori, K., Emoto, M., Inaba, M., 2012. Fetuin-A and the cardiovascular system. Adv. Clin. Chem. 56, 175–195.

Moschen, A.R., Adolph, T.E., Gerner, R.R., Wieser, V., Tilg, H., 2017. Lipocalin-2: a master mediator of intestinal and metabolic inflammation,. Trends Endocrinol. Metabol. 28, 388–397.

Mota, M., Banini, B.A., Cazanave, S.C., Sanyal, A.J., 2016. Molecular mechanisms of lipotoxicity and glucotoxicity in nonalcoholic fatty liver disease. Metabolism 65, 1049–1061.

Nandipati, K.C., Subramanian, S., Agrawal, D.K., 2017. Protein kinases: mechanisms and downstream targets in inflammation-mediated obesity and insulin resistance. Mol. Cell. Biochem. 426, 27–45.

Nielsen, A.R., Erikstrup, C., Johansen, J.S., Fischer, C.P., Plomgaard, P., Krogh-Madsen, R., Taudorf, S., Lindegaard, B., Pedersen, B.K., 2008. Plasma YKL-40: a BMI-independent marker of type 2 diabetes. Diabetes 57, 3078–3082.

Ofei, F., Hurel, S., Newkirk, J., Sopwith, M., Taylor, R., 1996. Effects of an engineered human anti-TNF-alpha antibody (CDP571) on insulin sensitivity and glycemic control in patients with NIDDM. Diabetes 45, 881–885.

Oh, K.J., Lee, D.S., Kim, W.K., Han, B.S., Lee, S.C., Bae, K.H., 2016. Metabolic adaptation in obesity and type II diabetes: myokines, adipokines and hepatokines. Int. J. Mol. Sci. 18.

Osborn, O., Brownell, S.E., Sanchez-Alavez, M., Salomon, D., Gram, H., Bartfai, T., 2008. Treatment with an Interleukin 1 beta antibody improves glycemic control in diet-induced obesity. Cytokine 44, 141–148.

Ouchi, N., Parker, J.L., Lugus, J.J., Walsh, K., 2011. Adipokines in inflammation and metabolic disease. Nat. Rev. Immunol. 11, 85–97.

Ozcan, U., Cao, Q., Yilmaz, E., Lee, A.H., Iwakoshi, N.N., Ozdelen, E., Tuncman, G., Gorgun, C., Glimcher, L.H., Hotamisligil, G.S., 2004. Endoplasmic reticulum stress links obesity, insulin action, and type 2 diabetes. Science 306, 457–461.

Panera, N., Della Corte, C., Crudele, A., Stronati, L., Nobili, V., Alisi, A., 2016. Recent advances in understanding the role of adipocytokines during non-alcoholic fatty liver disease pathogenesis and their link with hepatokines. Expet Rev. Gastroenterol. Hepatol. 10, 393–403.

Paquot, N., Castillo, M.J., Lefebvre, P.J., Scheen, A.J., 2000. No increased insulin sensitivity after a single intravenous administration of a recombinant human tumor necrosis factor receptor: Fc fusion protein in obese insulin-resistant patients. J. Clin. Endocrinol. Metab. 85, 1316–1319.

Paulsen, G., Mikkelsen, U.R., Raastad, T., Peake, J.M., 2012. Leucocytes, cytokines and satellite cells: what role do they play in muscle damage and regeneration following eccentric exercise? Exerc. Immunol. Rev. 18, 42–97.

Pedersen, B.K., Febbraio, M.A., 2008. Muscle as an endocrine organ: focus on muscle-derived interleukin-6. Physiol. Rev. 88, 1379–1406.

Pedersen, B.K., Saltin, B., 2006. Evidence for prescribing exercise as therapy in chronic disease. Scand. J. Med. Sci. Sports 16 (Suppl. 1), 3–63.

Pender, C., Goldfine, I.D., Kulp, J.L., Tanner, C.J., Maddux, B.A., MacDonald, K.G., Houmard, J.A., Youngren, J.F., 2005. Analysis of insulin-stimulated insulin receptor activation and glucose transport in cultured skeletal muscle cells from obese subjects. Metabolism 54, 598–603.

Petersen, A.M., Pedersen, B.K., 2005. The anti-inflammatory effect of exercise. J. Appl. Physiol. 98, 1154–1162.

Petersen, K.F., Dufour, S., Befroy, D., Garcia, R., Shulman, G.I., 2004. Impaired mitochondrial activity in the insulin-resistant offspring of patients with type 2 diabetes. N. Engl. J. Med. 350, 664–671.

Pillon, N.J., Bilan, P.J., Fink, L.N., Klip, A., 2013. Cross-talk between skeletal muscle and immune cells: muscle-derived mediators and metabolic implications. Am. J. Physiol. Endocrinol. Metab. 304, E453–E465.

Place, D.E., Kanneganti, T.D., 2017. Recent advances in inflammasome biology. Curr. Opin. Immunol. 50, 32–38.

Planavila, A., Redondo, I., Hondares, E., Vinciguerra, M., Munts, C., Iglesias, R., Gabrielli, L.A., Sitges, M., Giralt, M., van Bilsen, M., Villarroya, F., 2013. Fibroblast growth factor 21 protects against cardiac hypertrophy in mice. Nat. Commun. 4, 2019.

Radaelli, M.G., Martucci, F., Perra, S., Accornero, S., Castoldi, G., Lattuada, G., Manzoni, G., Perseghin, G., 2017. NAFLD/NASH in patients with type 2 diabetes and related treatment options. J. Endocrinol. Invest.

Rasouli, N., 2016. Adipose tissue hypoxia and insulin resistance. J. Invest. Med. 64, 830−832.

Rehman, K., Akash, M.S., 2016. Mechanisms of inflammatory responses and development of insulin resistance: how are they interlinked? J. Biomed. Sci. 23, 87.

Revollo, J.R., Korner, A., Mills, K.F., Satoh, A., Wang, T., Garten, A., Dasgupta, B., Sasaki, Y., Wolberger, C., Townsend, R.R., Milbrandt, J., Kiess, W., Imai, S., 2007. Nampt/PBEF/Visfatin regulates insulin secretion in beta cells as a systemic NAD biosynthetic enzyme. Cell Metabol. 6, 363−375.

Ridker, P.M., Howard, C.P., Walter, V., Everett, B., Libby, P., Hensen, J., Thuren, T., Cantos Pilot Investigative Group, 2012. Effects of interleukin-1beta inhibition with canakinumab on hemoglobin A1c, lipids, C-reactive protein, interleukin-6, and fibrinogen: a phase IIb randomized, placebo-controlled trial. Circulation 126, 2739−2748.

Rodriguez, A., Becerril, S., Ezquerro, S., Mendez-Gimenez, L., Fruhbeck, G., 2017. Crosstalk between adipokines and myokines in fat browning. Acta Physiol. 219, 362−381.

Rohrborn, D., Wronkowitz, N., Eckel, J., 2015. DPP4 in diabetes. Front. Immunol. 6, 386.

Romacho, T., Sanchez-Ferrer, C.F., Peiro, C., 2013. Visfatin/Nampt: an adipokine with cardiovascular impact. Mediat. Inflamm. 2013, 946427.

Romacho, T., Vallejo, S., Villalobos, L.A., Wronkowitz, N., Indrakusuma, I., Sell, H., Eckel, J., Sanchez-Ferrer, C.F., Peiro, C., 2016. Soluble dipeptidyl peptidase-4 induces microvascular endothelial dysfunction through proteinase-activated receptor-2 and thromboxane A2 release. J. Hypertens. 34, 869−876.

Roshanzamir, F., Miraghajani, M., Rouhani, M.H., Mansourian, M., Ghiasvand, R., Safavi, S.M., 2017. The association between circulating fetuin-A levels and type 2 diabetes mellitus risk: systematic review and meta-analysis of observational studies. J. Endocrinol. Invest.

Rui, L., Aguirre, V., Kim, J.K., Shulman, G.I., Lee, A., Corbould, A., Dunaif, A., White, M.F., 2001. Insulin/IGF-1 and TNF-alpha stimulate phosphorylation of IRS-1 at inhibitory Ser307 via distinct pathways. J. Clin. Invest. 107, 181−189.

Ruscica, M., Baragetti, A., Catapano, A.L., Norata, G.D., 2017. Translating the biology of adipokines in atherosclerosis and cardiovascular diseases: gaps and open questions. Nutr. Metabol. Cardiovasc. Dis. 27, 379−395.

Ryan, S., 2017. Adipose tissue inflammation by intermittent hypoxia: mechanistic link between obstructive sleep apnoea and metabolic dysfunction. J. Physiol. 595, 2423−2430.

Salska, A., Chizynski, K., 2016. Apelin - a potential target in the diagnosis and treatment of the diseases of civilization. Acta Cardiol. 71, 505−517.

Sartipy, P., Loskutoff, D.J., 2003. Monocyte chemoattractant protein 1 in obesity and insulin resistance. Proc. Natl. Acad. Sci. U. S. A. 100, 7265−7270.

Sattin, M., Towheed, T., 2016. The effect of TNFalpha-inhibitors on cardiovascular events in patients with rheumatoid arthritis: an updated systematic review of the literature. Curr. Rheumatol. Rev. 12, 208−222.

Schrauwen, P., Hesselink, M.K., 2004. Oxidative capacity, lipotoxicity, and mitochondrial damage in type 2 diabetes. Diabetes 53, 1412−1417.

Sell, H., Bluher, M., Kloting, N., Schlich, R., Willems, M., Ruppe, F., Knoefel, W.T., Dietrich, A., Fielding, B.A., Arner, P., Frayn, K.N., Eckel, J., 2013. Adipose dipeptidyl peptidase-4 and obesity: correlation with insulin resistance and depot-specific release from adipose tissue in vivo and in vitro. Diabetes Care 36, 4083−4090.

Sell, H., Habich, C., Eckel, J., 2012. Adaptive immunity in obesity and insulin resistance. Nat. Rev. Endocrinol. 8, 709−716.

Shah, R., Hinkle, C.C., Ferguson, J.F., Mehta, N.N., Li, M., Qu, L., Lu, Y., Putt, M.E., Ahima, R.S., Reilly, M.P., 2011a. Fractalkine is a novel human adipochemokine associated with type 2 diabetes. Diabetes 60, 1512−1518.

Shah, Z., Kampfrath, T., Deiuliis, J.A., Zhong, J., Pineda, C., Ying, Z., Xu, X., Lu, B., Moffatt-Bruce, S., Durairaj, R., Sun, Q., Mihai, G., Maiseyeu, A., Rajagopalan, S., 2011b. Long-term dipeptidyl-peptidase 4 inhibition reduces atherosclerosis and inflammation via effects on monocyte recruitment and chemotaxis. Circulation 124, 2338−2349.

Shah, Z., Pineda, C., Kampfrath, T., Maiseyeu, A., Ying, Z., Racoma, I., Deiuliis, J., Xu, X., Sun, Q., Moffatt-Bruce, S., Villamena, F., Rajagopalan, S., 2011c. Acute DPP-4 inhibition modulates vascular tone through GLP-1 independent pathways. Vasc. Pharmacol. 55, 2−9.

Shirakawa, J., De Jesus, D.F., Kulkarni, R.N., 2017. Exploring inter-organ crosstalk to uncover mechanisms that regulate beta-cell function and mass. Eur. J. Clin. Nutr. 71, 896−903.

Shoelson, S.E., Lee, J., Yuan, M., 2003. Inflammation and the IKK beta/I kappa B/NF-kappa B axis in obesity- and diet-induced insulin resistance. Int. J. Obes. Relat. Metab. Disord. 27 (Suppl. 3), S49−S52.

Sigal, R.J., Kenny, G.P., Boule, N.G., Wells, G.A., Prudhomme, D., Fortier, M., Reid, R.D., Tulloch, H., Coyle, D., Phillips, P., Jennings, A., Jaffey, J., 2007. Effects of aerobic training, resistance training, or both on glycemic control in type 2 diabetes: a randomized trial. Ann. Intern. Med. 147, 357−369.

Skurk, T., Alberti-Huber, C., Herder, C., Hauner, H., 2007. Relationship between adipocyte size and adipokine expression and secretion. J. Clin. Endocrinol. Metab. 92, 1023−1033.

So, B., Kim, H.J., Kim, J., Song, W., 2014. Exercise-induced myokines in health and metabolic diseases. Integr Med Res 3, 172−179.

Starkie, R., Ostrowski, S.R., Jauffred, S., Febbraio, M., Pedersen, B.K., 2003. Exercise and IL-6 infusion inhibit endotoxin-induced TNF-alpha production in humans. FASEB. J. 17, 884−886.

Stefan, N., Fritsche, A., Weikert, C., Boeing, H., Joost, H.G., Haring, H.U., Schulze, M.B., 2008a. Plasma fetuin-A levels and the risk of type 2 diabetes. Diabetes 57, 2762−2767.

Stefan, N., Haring, H.U., 2013. The role of hepatokines in metabolism. Nat. Rev. Endocrinol. 9, 144−152.

Stefan, N., Haring, H.U., Schulze, M.B., 2008b. Association of fetuin-A level and diabetes risk. J. Am. Med. Assoc. 300, 2247 author reply 47−48.

Stern, J.H., Rutkowski, J.M., Scherer, P.E., 2016. Adiponectin, leptin, and fatty acids in the maintenance of metabolic homeostasis through adipose tissue crosstalk. Cell Metabol. 23, 770−784.

Stienstra, R., van Diepen, J.A., Tack, C.J., Zaki, M.H., van de Veerdonk, F.L., Perera, D., Neale, G.A., Hooiveld, G.J., Hijmans, A., Vroegrijk, I., van den Berg, S., Romijn, J., Rensen, P.C., Joosten, L.A., Netea, M.G., Kanneganti, T.D., 2011. Inflammasome is a central player in the induction of obesity and insulin resistance. Proc. Natl. Acad. Sci. U. S. A. 108, 15324−15329.

Ta, N.N., Li, Y., Schuyler, C.A., Lopes-Virella, M.F., Huang, Y., 2010. DPP-4 (CD26) inhibitor alogliptin inhibits TLR4-mediated ERK activation and ERK-dependent MMP-1 expression by U937 histiocytes. Atherosclerosis 213, 429−435.

Tan, B.K., Adya, R., Randeva, H.S., 2010. Omentin: a novel link between inflammation, diabesity, and cardiovascular disease. Trends Cardiovasc. Med. 20, 143−148.

Taube, A., Schlich, R., Sell, H., Eckardt, K., Eckel, J., 2012. Inflammation and metabolic dysfunction: links to cardiovascular diseases. Am. J. Physiol. Heart Circ. Physiol. 302, H2148−H2165.

Tilg, H., Moschen, A.R., Roden, M., 2017. NAFLD and diabetes mellitus. Nat. Rev. Gastro-enterol. Hepatol. 14, 32−42.

Toldo, S., Abbate, A., 2017. The NLRP3 inflammasome in acute myocardial infarction. Nat. Rev. Cardiol.

Trayhurn, P., 2013. Hypoxia and adipose tissue function and dysfunction in obesity. Physiol. Rev. 93, 1−21.

Trayhurn, P., 2014. Hypoxia and adipocyte physiology: implications for adipose tissue dysfunction in obesity. Annu. Rev. Nutr. 34, 207−236.

Trayhurn, P., Wang, B., Wood, I.S., 2008a. Hypoxia and the endocrine and signalling role of white adipose tissue. Arch. Physiol. Biochem. 114, 267−276.

Trayhurn, P., Wang, B., Wood, I.S., 2008b. Hypoxia in adipose tissue: a basis for the dysre-gulation of tissue function in obesity? Br. J. Nutr. 100, 227−235.

Tsochatzis, E., Papatheodoridis, G.V., Archimandritis, A.J., 2006. The evolving role of leptin and adiponectin in chronic liver diseases. Am. J. Gastroenterol. 101, 2629−2640.

Ueda, S.Y., Yoshikawa, T., Katsura, Y., Usui, T., Fujimoto, S., 2009a. Comparable effects of moderate intensity exercise on changes in anorectic gut hormone levels and energy intake to high intensity exercise. J. Endocrinol. 203, 357−364.

Ueda, S.Y., Yoshikawa, T., Katsura, Y., Usui, T., Nakao, H., Fujimoto, S., 2009b. Changes in gut hormone levels and negative energy balance during aerobic exercise in obese young males. J. Endocrinol. 201, 151−159.

van der Poorten, D., Milner, K.L., Hui, J., Hodge, A., Trenell, M.I., Kench, J.G., London, R., Peduto, T., Chisholm, D.J., George, J., 2008. Visceral fat: a key mediator of steatohepatitis in metabolic liver disease. Hepatology 48, 449−457.

Vandanmagsar, B., Youm, Y.H., Ravussin, A., Galgani, J.E., Stadler, K., Mynatt, R.L., Ravussin, E., Stephens, J.M., Dixit, V.D., 2011. The NLRP3 inflammasome instigates obesity-induced inflammation and insulin resistance. Nat. Med. 17, 179−188.

Vane, J., Botting, R., 1987. Inflammation and the mechanism of action of anti-inflammatory drugs. FASEB. J. 1, 89−96.

Vashist, S.K., Schneider, E.M., Venkatesh, A.G., Luong, J.H.T., 2017. Emerging human fetuin a assays for biomedical diagnostics. Trends Biotechnol. 35, 407−421.

Verma, S., Hussain, M.E., 2017. Obesity and diabetes: an update. Diabetes Metab Syndr 11, 73−79.

Vernon, G., Baranova, A., Younossi, Z.M., 2011. Systematic review: the epidemiology and natural history of non-alcoholic fatty liver disease and non-alcoholic steatohepatitis in adults. Aliment. Pharmacol. Ther. 34, 274−285.

Wallenius, V., Wallenius, K., Ahren, B., Rudling, M., Carlsten, H., Dickson, S.L., Ohlsson, C., Jansson, J.O., 2002. Interleukin-6-deficient mice develop mature-onset obesity. Nat. Med. 8, 75−79.

Wang, S., Yang, X., 2017. Inter-organ regulation of adipose tissue browning. Cell. Mol. Life Sci. 74, 1765−1776.

Wang, Y., 2012. Small lipid-binding proteins in regulating endothelial and vascular functions: focusing on adipocyte fatty acid binding protein and lipocalin-2. Br. J. Pharmacol. 165, 603−621.

Weigert, C., Lehmann, R., Hartwig, S., Lehr, S., 2014. The secretome of the working human skeletal muscle—a promising opportunity to combat the metabolic disaster? Proteonomics Clin. Appl. 8, 5−18.

Weikert, C., Stefan, N., Schulze, M.B., Pischon, T., Berger, K., Joost, H.G., Haring, H.U., Boeing, H., Fritsche, A., 2008. Plasma fetuin-a levels and the risk of myocardial infarction and ischemic stroke. Circulation 118, 2555−2562.

Weisberg, S.P., McCann, D., Desai, M., Rosenbaum, M., Leibel, R.L., Ferrante Jr., A.W., 2003. Obesity is associated with macrophage accumulation in adipose tissue. J. Clin. Invest. 112, 1796−1808.

Wellen, K.E., Hotamisligil, G.S., 2005. Inflammation, stress, and diabetes. J. Clin. Invest. 115, 1111−1119.

Wen, H., Gris, D., Lei, Y., Jha, S., Zhang, L., Huang, M.T., Brickey, W.J., Ting, J.P., 2011. Fatty acid-induced NLRP3-ASC inflammasome activation interferes with insulin signaling. Nat. Immunol. 12, 408−415.

Wood, I.S., de Heredia, F.P., Wang, B., Trayhurn, P., 2009. Cellular hypoxia and adipose tissue dysfunction in obesity. Proc. Nutr. Soc. 68, 370−377.

Wronkowitz, N., Gorgens, S.W., Romacho, T., Villalobos, L.A., Sanchez-Ferrer, C.F., Peiro, C., Sell, H., Eckel, J., 2014a. Soluble DPP4 induces inflammation and proliferation of human smooth muscle cells via protease-activated receptor 2. Biochim. Biophys. Acta 1842, 1613−1621.

Wronkowitz, N., Romacho, T., Sell, H., Eckel, J., 2014b. Adipose tissue dysfunction and inflammation in cardiovascular disease. Front. Horm. Res. 43, 79−92.

Wu, H., Ballantyne, C.M., 2017. Skeletal muscle inflammation and insulin resistance in obesity. J. Clin. Invest. 127, 43−54.

Wu, L., Chen, L., Li, L., 2017a. Apelin/APJ system: a novel promising therapy target for pathological angiogenesis,. Clin. Chim. Acta 466, 78−84.

Wu, Y., Wang, X., Zhou, X., Cheng, B., Li, G., Bai, B., 2017b. Temporal expression of apelin/apelin receptor in ischemic stroke and its therapeutic potential. Front. Mol. Neurosci. 10, 1.

Xu, H., Barnes, G.T., Yang, Q., Tan, G., Yang, D., Chou, C.J., Sole, J., Nichols, A., Ross, J.S., Tartaglia, L.A., Chen, H., 2003. Chronic inflammation in fat plays a crucial role in the development of obesity-related insulin resistance. J. Clin. Invest. 112, 1821−1830.

Yamaguchi, S., Yoshino, J., 2017. Adipose tissue NAD(+) biology in obesity and insulin resistance: from mechanism to therapy. Bioessays 39.

Yamawaki, H., 2011. Vascular effects of novel adipocytokines: focus on vascular contractility and inflammatory responses. Biol. Pharm. Bull. 34, 307−310.

Yang, Q., Graham, T.E., Mody, N., Preitner, F., Peroni, O.D., Zabolotny, J.M., Kotani, K., Quadro, L., Kahn, B.B., 2005. Serum retinol binding protein 4 contributes to insulin resistance in obesity and type 2 diabetes. Nature 436, 356−362.

Yang, S.J., Hwang, S.Y., Choi, H.Y., Yoo, H.J., Seo, J.A., Kim, S.G., Kim, N.H., Baik, S.H., Choi, D.S., Choi, K.M., 2011. Serum selenoprotein P levels in patients with type 2 diabetes and prediabetes: implications for insulin resistance, inflammation, and atherosclerosis. J. Clin. Endocrinol. Metab. 96, E1325−E1329.

Yang, Y., Lv, S.Y., Ye, W., Zhang, L., 2016. Apelin/APJ system and cancer. Clin. Chim. Acta 457, 112−116.

Yazici, D., Sezer, H., 2017. Insulin resistance, obesity and lipotoxicity. Adv. Exp. Med. Biol. 960, 277−304.

Ye, J., 2009. Emerging role of adipose tissue hypoxia in obesity and insulin resistance. Int. J. Obes. 33, 54−66.

Yoo, H.J., Choi, K.M., 2015. Hepatokines as a link between obesity and cardiovascular diseases. Diabetes Metab. J 39, 10−15.

Younossi, Z., Anstee, Q.M., Marietti, M., Hardy, T., Henry, L., Eslam, M., George, J., Bugianesi, E., 2017. Global burden of NAFLD and NASH: trends, predictions, risk factors and prevention. Nat. Rev. Gastroenterol. Hepatol.

Yu, H., Xia, F., Lam, K.S., Wang, Y., Bao, Y., Zhang, J., Gu, Y., Zhou, P., Lu, J., Jia, W., Xu, A., 2011. Circadian rhythm of circulating fibroblast growth factor 21 is related to diurnal changes in fatty acids in humans. Clin. Chem. 57, 691−700.

Zhang, X., Ji, X., Wang, Q., Li, J.Z., 2017. New insight into inter-organ crosstalk contributing to the pathogenesis of non-alcoholic fatty liver disease (NAFLD). Protein Cell.

Zhao, L., Ackerman, S.L., 2006. Endoplasmic reticulum stress in health and disease. Curr. Opin. Cell Biol. 18, 444−452.

Zhong, J., Rao, X., Rajagopalan, S., 2013. An emerging role of dipeptidyl peptidase 4 (DPP4) beyond glucose control: potential implications in cardiovascular disease. Atherosclerosis 226, 305−314.

Zhou, Q., Cao, J., Chen, L., 2016. Apelin/APJ system: a novel therapeutic target for oxidative stress-related inflammatory diseases (review). Int. J. Mol. Med. 37, 1159−1169.

Zierath, J.R., Krook, A., Wallberg-Henriksson, H., 2000. Insulin action and insulin resistance in human skeletal muscle. Diabetologia 43, 821−835.

Technical Annex

CHAPTER OUTLINE

The Cellular Secretome and Organ Crosstalk. https://doi.org/10.1016/B978-0-12-809518-8.00006-4

155

In this annex a compendium of laboratory protocols is presented comprising a variety of methods used for the analysis of crosstalk between different tissues and cells. All protocols have been published (see list of references) and/or are described in detail in a number of master and doctoral theses. These methods were used in the author's laboratory for many years, and they should be considered as a good starting point when planning to perform experimental work in this field of research. My special thanks go to all my former collaborators and students in the Paul-Langerhans-Group for Integrative Physiology, who prepared and contributed to this valuable collection of methods.

CELL ISOLATION AND CULTURE PROTOCOLS

We present here a collection of protocols for the culture of adipocytes, myocytes, cardiomyocytes, vascular cells, and pancreatic β-cells. In many cases, the protocols use primary human cells. These cells are a rich source for preparation of conditioned media that can be used for crosstalk analysis.

ISOLATION OF PREADIPOCYTES

Materials	Distributor/ Manufacturer	Cat. No.	Storage (°C)
Collagenase NB 4G Proved Grade	Serva	1746502	4
BSA (bovine serum albumin), Fraction V, fatty acid–free	Roth	0052.3	4
10× PBS (phosphate buffer saline) (without Ca/Mg)	PAA		4
Other chemicals from Sigma			
150 µm mesh	Fisher		
Plastic filter with 70 µm mesh	Falcon		

Preparation of Solutions and Buffers

Erythrocyte Lysis Buffer

155 mM NH_4Cl	8.29 g/L	
5.7 mM K_2HPO_4	0.99 g/L	
0.1 mM	0.04 g/L	pH 7.3 and sterilize

Collagenase Solution
    1× DPBS without Ca/Mg
    2% BSA, fatty acid–free
    250 U/mL collagenase
    pH 7.4

Procedure

1. Prepare fat pieces without connective tissue and vessels and store them in adipocyte basal medium overnight at 4°C (if necessary).
2. Weigh 15 g of fat in every 50-mL Falcon tube and mince the fat in very small pieces of approximately 10 mg.

3. Prepare sterile collagenase solution (30 mL per Falcon).
4. Mix fat pieces with collagenase solution and incubate for 90 min at 37°C (water shaker).
5. Centrifuge for 10 min at 1100 rpm and throw away watery phase and swimming fat (see picture).

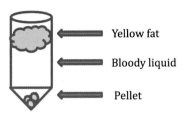

6. Add 10 mL erythrocyte lysis buffer to pellet, resuspend, and filter through glass filter with 150 μm mesh.
7. Centrifuge for 10 min at 1100 rpm and resolve pellet in adipocyte basal medium.
8. Filter through 70 μm plastic filter and count cells.
9. Seed in appropriate dishes.

Seeding Density

175.000 cells per 12-well cavity (2.1 Mio per plate)
350.000 cells per 6-well cavity (2.1 Mio per plate)
1 Mio cells per T25
3.5 Mio cells per T75

PREADIPOCYTE PROLIFERATION

Materials	Distributor/Manufacturer	Cat. No.	Storage (°C)
Insulin, human recombinant	Sigma	91077C	−20
rh FGF-b	Immunotools	11343625	−20
rh EGF	Immunotools	11343407	−20
FCS	Gibco	10270	−20
Gentamycin 10 mg/mL	Invitrogen	15710049	4
Trypsin 0.05%/EDTA (1×)	Invitrogen	25300-054	−20

Preparation of Solutions and Buffers
Stocks

Human insulin stock:

- Weigh 5 mg of insulin per 1 mL of 5 mM HCl (made from Titrisolve).
- Freeze in aliquots of 500 μL to prepare 50 mL of medium or aliquots of 5 mL to prepare 500 mL of medium.

Human recombinant EGF:

- Dissolve in 1 mL sterile H_2O containing 0.1% BSA as a carrier protein (stock of 500 µg/mL).
- Freeze in aliquots of not bigger than 10 µL.
- Use 2 µL to prepare 100 mL of medium or 10 µL to prepare 500 mL of medium.

Human recombinant bFGF:

- Dissolve in 500 µL sterile H_2O containing 0.1% BSA as a carrier protein (stock of 100 µg/mL).
- Freeze in aliquots of not bigger than 5 µL.
- Use 1 µL to prepare 100 mL of medium or 5 µL to prepare 500 mL of medium.

Proliferation Medium

For 100 mL, use the following composition:

Components	Final Concentration
• 1 mL of human insulin stock of 870 µM	8.7 µM
• 2.5 mL FCS	2.5% (v/v)
• 2 µL EGF	10 ng/mL
• 1 µL bFGF	1 ng/mL
• 500 µL gentamicin	0.5% (v/v)

Procedure

1. **Freshly isolated adipocytes** are seeded at a density of approximately 5000 cells per cm^2 (125,000 cells per T25; 375,000 cells per T75; 38,000 per 6-well cavity) **in growth medium** (see protocol for adipocyte differentiation).
2. After 1 day, **change** medium **to proliferation medium** and do medium changes every 2−3 days.
3. Subculture/**passage** the cells when subconfluence is reached and seed at the same density as above.
   - Wash cells with warm PBS.
   - Add trypsin for about 5 min at 37°C and stop digestion by adding growth medium (use 2−3 mL of trypsin per T75 and add twice the volume of growth medium).
   - Centrifuge the cell dispension at 1200 rpm for 5 min and resuspend the pellet in proliferation medium for cell counting.

dish	ml Trypsin	cell no. seeded	In x ml medium	days until confluence
T175	5 ml	0,8 − 1 Mio	20 ml	6-7 days
T75	2.5 ml	500,000	10 ml	

4. Freeze preadipocytes after 1 passage ($\rightarrow$ P2) in cryomedium from PromoCell.
  - Detach cells according to the subculture procedure above.
  - Count cells.
  - Centrifuge cell suspension at 1200 rpm for 5 min and resuspend cell pellet in an appropriate amount of cryomedium:
    - 3 Mio cells/mL cryomedium (per cryotube).
    - Label tubes, for example, F382 P2, 3 Mio/mL, date, name.

5. Defrost cells/seed for experiment.
  - Resuspend one cryovial (containing 3 Mio cells) in 36 mL growth medium $\rightarrow$ sufficient for 3 $\times$ 6-well plates (170.000 per 6-well cavity).
  - Change medium as soon as the cells are attached, at the latest the next day.
  - Differentiation can be started after 3–5 days.

ISOLATION OF MURINE PREADIPOCYTES

Materials	Distributor/ Manufacturer	Cat. No.	Storage (°C)
Collagenase NB 4G Proved Grade	Serva	1746502	4
BSA, Fraction V, fatty acid–free	Roth	0052.3	4
10× PBS (without Ca/Mg)	PAA		4
Plastic filter with 70 µm mesh	Falcon		
Alpha MEM	Gibco	22561-021	4
Biotin	Sigma	B4639	4
DCS	Invitrogen	16030074	4
StemPro Accutase	Invitrogen	A11105-01	4

Preparation of Solutions and Buffers

Erythrocyte Lysis Buffer
     155 mM NH_4Cl 8.29 g/L
     5.7 mM K_2HPO_4 0.99 g/L
     0.1 mM   0.04 g/L   pH 7.3 and sterilize
Collagenase Solution
     1× DPBS without Ca/Mg
     2% BSA, fatty acid–free
     250 U/mL collagenase $\rightarrow$ 1 mg/mL per 2 mL adipose tissue (AT)
     pH 7.4, filter sterile!
Complete Medium
     439 mL Alpha MEM
     5 Ll antibiotic-antimycotic
     5 Ll biotin stock $\rightarrow$ final concentration 8 µM
     500 µL ascorbic acid (stock 100 mM) $\rightarrow$ final concentration 100 µM
     500 µL pantothenic acid $\rightarrow$ final concentration 18 µM
     50 mL DCS $\rightarrow$ final concentration 10%

Procedure

Isolation

1. Prepare Falcon tubes with sterile PBS → weigh.
2. Prepare AT from designated fat pad of mice → pool depots of at least 3 mice.
3. Prepare collagenase solution → ~10 mL per Falcon tube.
4. Transfer AT from Falcon into new Falcon and add collagenase.
5. Mince tissue with sterile scissors.
6. Incubate at 37°C in a water bath with shaking for 45−60 min (until fat totally digested).
7. Centrifuge for 10 min at 1100 rpm, room temperature (RT).
8. Resuspend pellet in 3 mL erythrocyte lysis buffer and incubate for 10min.
9. Centrifuge for 10 min at 1100 rpm RT.
10. Resuspend pellet in ~3 mL complete medium.
11. Filter through 70 μm mesh.
12. Count cells → seed in appropriate density.
13. Next day, wash 2× with sterile PBS and add new medium.
14. Change medium every 2−3 days until 80% confluent → split cells.

Splitting

1. Wash cells with sterile PBS.
2. Add appropriate volume of accutase (2 mL for T25, 5 mL for T75).
3. When all cells are in solution, stop reaction by adding same volume of complete medium.
4. Centrifuge for 5 min at 1100 rpm RT.
5. Count cells in Neubauer chamber.
6. Seed in new flask → freeze with 1 mL Cryo-SFM (at least 2 Mio/mL).

Differentiation

Differentiation Medium (for 50 mL)
- 48.6 mL complete medium
- 1.32 mL insulin stock → final 861 nM
- 33.3 μL dexamethasone (1:100 predilution of stock) → final 33.3 nM
- 15.5 μL T3 → final 3.1 nM
- 1.8 μL rosiglitazone → final 1 μM
- 25 μL transferrin → final 10 μg/mL

Starvation Medium
- Alpha MEM + 1% antibioticum-antimycoticum

Procedure
1. Start differentiation when cells are 80%−90% confluent by adding 2 mL differentiation medium per well.
2. Change every 2−3 days.
3. Culture 9−14 days.

MATURE ADIPOCYTE ISOLATION

Materials	Distributor/ Manufacturer	Cat. No.	Storage (°C)
Collagenase NB 4G Proved Grade	Serva	1746502	4
BSA, Fraction V, fatty acid—free	Roth	0052.3	4
10× PBS (without Ca/Mg)	PAA		4
Other chemicals from Sigma			
Plastic filter with 70 μm mesh	Falcon		

Preparation of Solutions and Buffers
Collagenase Solution

> 1× DPBS without Ca/Mg
> 2% BSA, fatty acid—free
> 250 U/mL collagenase (1 mg/mL)
> pH 7.4

Procedure

- Prepare sterile collagenase solution (10 mL per Falcon tube).
- Mince AT and add collagenase solution.
- Incubate at 37°C 30—45 min in a shaking water bath until the solution is cloudy and the pieces are more or less digested.
- Centrifuge at 1100 rpm for 10 min RT.
- Prepare Falcon tubes with 10 mL PBS.

Mature Adipocytes

- Washing of mature adipocytes: Transfer upper mature adipocyte phase with cut 1 mL tips into PBS-Falcons and carefully invert Falcon several times.
- Centrifuge at 500 rpm for 5 min RT.
- Repeat washing step.
- Transfer ∼ 400 μL mature adipocytes (try to avoid transfer of too much PBS) into 2 mL Safe-Lock Eppendorf tube.
- Centrifuge in cooling centrifuge at 900 rpm for 5min and remove with an insulin syringe or a filter tip the remaining PBS → directly freeze in liquid nitrogen.

SVF (Stroma Cascular Fraction)

- Discard interphase after 10 min of centrifugation and dissolve pellet in an appropriate amount of PBS (according to size of the pellet 2—4 mL).
- Filter through a 70 μm plastic filter.
- Distribute filtered cell solution in 2—4 Safe-Lock tubes.
- Centrifuge in a cooling centrifuge at 1100 rpm, 5min, 4°C.

- Discard supernatant → directly freeze in liquid nitrogen.
- Afterward, isolate RNA or protein with the according buffers and homogenate protein with the use of a homogenizer under cooled conditions!

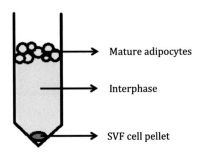

PREPARATION OF EXPLANTS FROM ADIPOSE TISSUE

Materials	Distributor/ Manufacturer	Cat. No.	Storage
Scalpel (nr 22, sterile)	VWR	233-0028	RT
PBS, no calcium, no magnesium	Gibco	14190-169	4°C or RT
Antibiotic-Antimycotic (100×)	Gibco	15240-096	−20°C
150 μm mesh or smaller	Fisher		

Preparation of Solutions and Buffers
PBS with AA-Mix (Antibiotic-Antimycotic [100×] Mix)

- Add 1 mL of AA-mix to 500 mL of PBS.

Procedure
1. Use AT as fresh as possible (if necessary store the fat in sterile medium or PBS).
2. Remove connective tissue and vessels.
3. Cut AT with two scalpels into small pieces of around 10 mg.
4. Wash explants three times in warm PBS with AA-mix (15 mL) using a cut blue pipette tip and centrifuge (1 min at 300 ×g) between the washing steps.
5. Remove liquid by putting explants on a sterile mesh and weight relatively dry bulk of explants before adding medium.
6. After optional culture for 24 h in medium, 100 mg of AT is used to generate 1 mL of conditioned medium for 24 h at 37°C and 5% CO_2.
7. Collect conditioned medium in a tube and store at −80°C.
8. If needed remove liquid of explants by putting them on a sterile mesh and freeze at −80°C until used for lysates.

CULTURE OF HUMAN ADIPOCYTES

Materials	Distributor/ Manufacturer	Cat. No.	Storage
Millipore Steritop GP Filter, 0.22 μm	Fisher	3422553	RT
Sterile syringe filter, 0.2 μm cellulose acetate membrane	VWR	514-0061	RT
DMEM:F12 (without $NaHCO_3$)	Invitrogen	42400-010	4°C
$NaHCO_3$	Merck	6329	RT
Biotin	Sigma	B-4639-500 mg	4°C
Pantothenic acid	Sigma	P5155-100 g	4°C
Human insulin	Sigma	I-2643	−20°C
T3	Sigma	T-6397	−20°C
Apo-Transferrin	Sigma	T 2036	−20°C
Gentamicin 10 mg/mL	Invitrogen	15710049	4°C
FCS	Gibco	10270	−20°C
Troglitazone	Sigma		−20°C
Antibiotic-Antimycotic (100×)	Gibco	15240-096	−20°C

Human Preadipocyte Media
Stocks
Human Insulin (33 μM)

> 1.9 mg + 10 mL 5 mM HCl
> Filter through sterile syringe filter 0.2 μm
> 1.5 mL per aliquot

Troglitazone (5 mM)

> 2.21 mg + 1 mL DMSO
> Filter through sterile syringe filter 0.2 μm
> 100 μL per aliquot

Transferrin (20 mg/mL)

> 100 mg in 5 mL sterile PBS
> 250 μL per aliquot

T3

> Stock solution: 100 mg T3 in 3 mL EtOH + 3 mL 1N NaOH
> Ready-to-use solution: 20 μM → dilute 1:25 with EtOH
> 50 μL per aliquot

Basal Medium

> 2× DMEM:F12 (without $NaHCO_3$)
> 2.25 g $NaHCO_3$
> 16 mg biotin
> 8 mg pantothenic acid

- Add 2 L Millipore water.
- Adjust pH to 7.4 (7.35–7.45).
- Use autoclaved 500 mL bottles and sterilize by filtration using Millipore Steritop GP Filter.
- Store at 4°C until use.

Growth Medium

> 500 mL basal medium
> 50 mL FCS
> 5.5 mL antibiotic-antimycotic (100×)

- Add all supplements to a bottle of basal medium.
- Store at 4°C.

Differentiation Medium

> 500 mL basal medium
> 3 mL human insulin (33 µM)
> 50 µL T3
> 1 mL cortisol
> 250 µL transferrin
> 2.5 mL gentamicin

- Add all supplements to a bottle of basal medium.
- Store at 4°C.

Differentiation + Troglitazone Medium

- Calculate needed medium for your cells and add 1 µL per mL troglitazone (stock 5 mM).
- Prepare freshly before use.

Procedure

Seeding

- Seed the isolated cells according to the preadipocyte isolation protocol in growth medium section.

Day After Seeding: Change Medium

- Change medium using growth medium.
- Change medium every 2–3 days using growth medium.
- Grow cells until they are confluent.

Day 0: Start Differentitation

- Add 1 µL per mL troglitazone to differentiation medium.
- Culture adipocytes for 3 days with troglitazone.

Day 3: Change to Differentiation Medium

- Use differentiation medium for cultivation of adipocytes.
- Change medium every 2–3 days.

Day 14: End of Differentiation

- Stimulate your cells with hSkMC Diff (−) medium.

DIFFERENTIATION OF PREADIPOCYTES (ALTERNATIVE METHOD)

Materials	Distributor/Manufacturer	Cat. No.	Storage
DMEM/F12	Invitrogen	42400-010	4°C
NaHCO$_3$	Merck	6329	RT
Biotin	Sigma	B-4639-500 mg	4°C
Pantothenate	Sigma	P5155-100 g	4°C
Human insulin	Sigma	I-2643	−20°C
Gentamycin	Invitrogen	15710049	4°C
FCS	Gibco	10270	−20°C
IBMX	Sigma	I5879_1G	4°C
Dexamethasone	Sigma	D4902-25 mg	4°C
0.2 μm sterile filters	VWR		
Troglitazone	Sigma		−20°C

Dexamethasone stock solution (5 mM)

Add 5 mg to 2.55 mL of DMSO, filter sterile, and prepare 50 μL aliquots to be stored at −20°C.

IBMX stock solution (494.82 mM)

Add 0.11 g of IBMX powder (sterile) to 1 mL of sterile DMSO, prepare 50 μL aliquots, and store at −20°C.

Human insulin stock solution (33 μM)

Add 1.9 mg into 10 mL of 5 mM HCl, mix well and filter sterile, prepare 250 μL aliquots, and store at −20°C.

- HCl 5 mM solution:
Take 62.5 μl from a 4 N HCl stock solution and add it to a final volume of 50 ml of Milli Q water.

Troglitazone (5 mM)

2.21 mg + 1 mL DMSO
Filter through sterile syringe filter 0.2 μm
50 μl per aliquot
Gentamicin commercial stock solution (10 mg/mL)
Growth adipocyte medium (check specific protocol for further details)

Basal adipocyte medium (Check specific protocol for further details)

- 500 mL DMEM/F12
- 33 μM biotin (Add 17 mg of biotin per 2 L medium)
- 17 μM pantothenate (Add 8 mg pantothenate per 2 L medium)

- Maintenance medium:

Add to basal adipocyte medium the following:	100 mL
• FCS (final concentration, 3%)	3 mL
• Human insulin (final concentration, 100 nM)	302.8 μL
• Dexamethasone (final concentration, 1 μM)	10 μL
• Gentamicin	500 μL

- Differentiation medium:
  Add to **50 mL** of maintenance medium the following:
- 20 μL of IBMX (0.2 mM).
- 2.5 μL of troglitazone (0.25 μM).
- Keep cells in the differentiation medium for a **whole week**, changing medium every 2–3 days, and then switch to maintenance.

Procedure

The preadipocytes are plated in growth adipocyte medium and changed every 2–3 days. Once the preadipocytes reach confluence, differentiation is induced with freshly prepared differentiation medium 2 mL per well (this medium can remain active for 1 week; do not prepare big amounts). The differentiation medium must stay on the cells for a total period of 1 week with 2–3 days' medium change. Once the initial differentiation week is over (day 7), the cells are switched to maintenance medium with a medium change (2 mL per well) every 2–3 days.

CULTURE OF PRIMARY HUMAN SKELETAL MUSCLE CELLS

Materials	Distributor/ Manufacturer	Cat. No.	Storage
Ham's F-12 Nutrient Mix, powder	Gibco	21700-026	4°C
Alpha MEM, powder	Gibco	11900-016	4°C
Antibiotic-Antimycotic (100×)	Gibco	15240-062	−20°C
Horse serum	Gibco	16050-122	−20°C
PBS, no calcium, no magnesium	Gibco	14190-169	4°C or RT
NaHCO$_3$ (cell culture grade)	AppliChem	A0384	RT
Supplement Pack for skeletal muscle cells	PromoCell	C-39360	−20°C
Millipore Steritop GP Filter	Millipore	3422553	RT
Detach Kit (HepesBSS, Trypsin, TNS)	PromoCell	41220	−20°C
Cyro-SFM	PromoCell	C-29912	4°C

Preparation of Solutions and Buffers

1. F12/Alpha MEM stock for growth medium, 2 L:

1 pack Ham's F-12 Nutrient Mix
1 pack Alpha MEM
3.3 g $NaHCO_3$

- Dissolve in 1.8 L distilled H_2O, adjust pH to 7.35, and add distilled H_2O to a total volume of 2 L.
- Sterilize by filtration using Millipore Steritop GP Filter and autoclaved 500 mL bottles.
- Store at 4°C until use.
- **To complete the growth medium, add one "Supplement Pack for skeletal muscle cells" and 1 mL antibiotic-antimycotic to 500 mL of F12/Alpha MEM.**

2. Alpha MEM stock for differentiation medium, 2 L:

2 packs Alpha MEM
4.4 g $NaHCO_3$

- Dissolve in 1.8 L distilled H_2O, adjust pH to 7.35, and add distilled H_2O to a total volume of 2 L.
- Sterilize by filtration using Millipore Steritop GP Filter and autoclaved 500 mL bottles.
- Store at 4°C until use.
- **To complete Diff (+) medium, add 10 mL horse serum and 1 mL antibiotic-antimycotic to 500 mL of Alpha MEM.**
- **To complete Diff (−) medium add 1 mL antibiotic-antimycotic to 500 mL of Alpha MEM.**

Procedure

1. Splitting the cells:

- Always use a separate bottle of growth medium when cultivating the cells for preparation of cell stocks.
- PromoCell delivers proliferating cells in T25 flask: After arrival of the cells supply them with fresh growth medium.
- Tebu Bio and Lonza deliver cryovials (>500,000 cells/vial): Store in liquid nitrogen until use.
  Prepare four T25 flasks with 7.5 mL growth medium.
  Thaw vial and dilute the cells in 10 mL medium.
  Add 2.5 mL cell suspension to 7.5 mL medium in a T25 flask.
- Allow them to reach 80% confluence.
- Remove the medium and rinse the cells 2× with HepesBSS (2.5 mL for T25; 5 mL for T175 flask).
- Remove HepesBSS and add trypsin (2.5 mL for T25; 5 mL for T175 flask); wait until the cells detach (gently tap the flask and check under the microscope).

- Add trypsin neutralization solution (TNS) (2.5 mL for T25; 5 mL for T175 flask) and collect the cell suspension in a 50 mL Falcon tube.
- Centrifuge for 5 min at 1300 ×g, remove supernatant, and add 1 mL growth medium.
- Carefully resuspend the cell pellet and count the cells: Mix, e.g., 10 µL cell suspension +90 µL growth medium +100 µL trypan blue; count the cells using a Neubauer chamber.
- Seed the cells in T175 flask for further cultivation (0.7−1 Mio cells/flask).
- Cells are frozen after the third trypsinization: For this, centrifuge again after counting (5 min at 1300 ×g) and resuspend the cell pellet in an appropriate volume of Cryo-SFM to get 2.4 Mio cells/mL.
- Aliquot the cells in cyrovials (1 mL) and freeze them in −80°C using a cooling box.
- Transfer the vials to liquid nitrogen the next day.

2. Seeding for experiments:
   - Thaw cryovial (2.4 Mio hSkMC/vial) in a water bath at 37°C.
   - Gently mix the cell suspension and transfer it to a Falcon tube containing an appropriate amount of growth medium (e.g., 48 mL growth medium to seed cells in four 6-well dishes).
   - Seed the cells according to your needs.
   - Change medium on the next day.
   - Allow them to reach 90% confluence, then switch to differentiation medium for 5 days.
   - For EPS experiments Diff (+) must be used.

CULTURE OF C2C12 MYOCYTES

Materials	Distributor/ Manufacturer	Cat. No.	Storage
DMEM (high glucose, pyruvate)	Gibco	41966	4°C
FCS	Gibco	10270106	−20°C
Antibiotic-Antimycotic (100×)	Gibco	15240-062	−20°C
Horse serum	Gibco	16050-122	−20°C
0.05% Trypsin-EDTA (1×)	Gibco	25300-054	4°C
PBS, no calcium, no magnesium	Gibco	14190-169	4°C or RT

Preparation of Solutions and Buffers

1. C2C12 growth medium:

   500 mL DMEM
   50 mL FCS
   1 mL antibiotic-antimycotic

   2. C2C12 Diff (+) medium:
      500 mL DMEM
      10 mL horse serum
      1 mL antibiotic-antimycotic
   3. C2C12 Diff (−) starvation medium:
      500 mL DMEM
      1 mL antibiotic-antimycotic

Procedure

   1. Thaw cryovial containing the cells in a water bath at 37°C.
   2. Mix cell suspension and transfer to a Falcon tube with 10 mL growth medium.
   3. Centrifuge for 2 min at 800 rpm and remove supernatant.
   4. Resuspend pellet in 10 mL growth medium and seed in a T75 flask.
   5. Change medium on the next day.
   6. Split the cells when they reach ∼60% confluence.
   7. To split the cells, remove medium and wash the cells with warm PBS.
   8. Remove PBS and add 2 mL 1× trypsin for 2−5 min at 37°C and check under the microscope.
   9. When the cells detach, add 15 mL growth medium and transfer the mixture to a Falcon tube.
   10. Centrifuge for 2 min at 800 rpm, remove supernatant, and add 2 mL growth medium.
   11. Count the cells and seed them in a T75 flask for further cultivation or, e.g., in 6-well dishes for experiments.

Differentiation of C2C12

1. Seed the cells in, e.g., 6-well dishes using growth medium; change medium every second day.
2. Allow them to reach 80% confluence, then switch to Diff (+) medium for 5 days.
3. At day 5, wash the cells 1× with warm PBS, then add Diff (−) medium over-night.
4. Perform the experiment.

CULTURE OF PRIMARY HUMAN SMOOTH MUSCLE CELLS

Materials	Distributor/ Manufacturer	Cat. No.	Storage
FCS	Gibco	10270	−20°C
SMC Medium	PromoCell	#C-22062	4°C
Gentamicin	Gibco	15710-049	RT
Fungizone	Invitrogen	15290-026	−4°C
Detach Kit (HepesBSS, Trypsin, TNS)	PromoCell	41220	−20°C
Cyro-SFM	PromoCell	C-29912	4°C

Preparation of Solutions and Buffers

1. SMC growth medium (DMEM, low glucose, Pyruvate, HEPES, und Glutamine)
   - 50 mL FCS
   - 2.5 mL gentamicin
   - 100 μL Fungizone
2. SMC starvation medium (DMEM, low glucose, Pyruvate, HEPES, und Glutamine)
   - 2.5 mL gentamicin
   - 100 μL Fungizone

Procedure

1. Seeding for experiments:
   - Seed 200,000 cells/well in growth medium (2 mL).
   - Culture for 24 h (until 50%–80% confluence).
   - Wash 1× with sterile PBS (2 mL).
   - Starvation for 24 h in starvation medium (2 mL).
2. Splitting the cells:
   - Always use a separate bottle of growth medium when cultivating the cells for preparation of cell stocks.
   - After arrival of the cells, supply them with fresh growth medium (PromoCell delivers proliferating cells in T25 flask).
   - Allow them to reach 80% confluence.
   - Remove the medium and rinse the cells 2× with HepesBSS (2.5 mL for T25; 5 mL for T175 flask).
   - Remove HepesBSS and add trypsin (2.5 mL for T25; 5 mL for T175 flask); wait until the cells detach (gently tap the flask; check under the microscope).
   - Add TNS (2.5 mL for T25; 5 mL for T175 flask) and collect the cell suspension in a 50-mL Falcon tube.
   - Centrifuge for 5 min at 1300 ×g, remove supernatant, and add 1 mL growth medium.
   - Carefully resuspend the cell pellet and count the cells: Mix, e.g., 10 μL cell suspension +90 μL growth medium + 100 μL trypan blue; count the cells using a Neubauer chamber.
   - Seed the cells in a T175 flask for further cultivation (750,000–800,000/flask).
   - Cells are frozen after the third trypsinization: For this, centrifuge again after counting (5 min at 1300 ×g) and resuspend the cell pellet in an appropriate volume of Cryo-SFM to get 2 Mio cells/mL.
   - Aliquot the cells in cyrovials (1–1.5 mL) and freeze them in −80°C using a cooling box.
   - Transfer the vials to liquid nitrogen the next day.

CULTURE OF HUMAN CORONARY ARTERY ENDOTHELIAL CELLS

Materials	Distributor/ Manufacturer	Cat. No.	Storage
FCS	Gibco	10270	−20°C
EC Basal Medium MV	PromoCell	#C-22220	4°C
Supplement Mix	PromoCell		−20°C
Gentamicin	Gibco	15710-049	4°C
Fungizone	Invitrogen	15290-026	−20°C
Trypsin	PromoCell	41220	−20°C/4°C
Cryo-SFM	PromoCell	C-29912	4°C
Antibiotic-Antimycotic (100×)	Gibco	15240-062	−20°C

Preparation of Solutions and Buffers

1. EC Medium MV 20% FCS (500 mL, growth medium)
   - 500 mL EC basal medium
   - 27.2 mL Supplement Mix
   - 81 mL FCS
   - 1 mL gentamicin
   - 100 µL Fungizone
2. EC Medium MV 5% FCS (500 mL, for treatments)
   - 500 mL EC basal medium
   - 27.2 mL FCS (heat-inactivated FCS—20 to 30 min at 56 degrees)
   - 1 mL gentamicin
   - 100 µL Fungizone
3. DMEM 10% FCS (500 mL, neutralization medium)
   - 500 mL Alpha DMEM medium
   - 50 mL FCS
   - 1 mL antibiotic-antimycotic

Procedure

1. Seeding:
   - Thaw one vial (1 Mio cells) and add required amount to 6-well plate (2 mL/well) or T75 flask (10 mL) in growth medium.
   - Culture until 90% confluence.
2. Splitting the cells:
   - Remove the medium and wash the cells 1× with PBS (5 mL for T75 flask).
   - Remove PBS and add a pulse of trypsin (2 mL for T75 flask); quickly remove trypsin.

- Add trypsin (1 mL for T75 flask), put the flask into the incubator, and wait until the cells detach (gently tap the flask; check under the microscope).
- Add trypsin neutralization medium (DMEM 10% FCS) (4 mL per mL trypsin) and collect the cell suspension in a 15 mL Falcon tube.
- Collect 10 μL cell suspension for counting. Count the cells using a Neubauer chamber, centrifuge for 7 min at 1200 ×g, remove supernatant, and add 5 mL growth medium.
- Carefully resuspend the cell pellet.
- Seed the cells in a T75 flask for further cultivation (250,000–500,000/flask; depending on demand) or for experiments in 6-well plates (>50,000 per well).

4. Freeze cells:
- Cells are frozen after the first or second passage: For this, resuspend the cell pellet in an appropriate volume of Cryo-SFM to get 1 Mio cells/mL.
- Aliquot the cells in cryovials (1–1.5 mL) and freeze them in −80°C using a cooling box for 24 h.
- Transfer the vials to liquid nitrogen the next day.

CULTURE OF HEPG2 LIVER CELLS

Materials	Distributor/ Manufacturer	Cat. No.	Storage
RPMI 1640	Gibco	21875-091	4°C
FCS	Gibco	10270106	−20°C
Antibiotic-Antimycotic (100×)	Gibco	15240-062	−20°C
0.05% Trypsin-EDTA (1×)	Gibco	25300-054	4°C
PBS, no calcium, no magnesium	Gibco	14190-169	4°C or RT

Preparation of Solutions and Buffers

HepG2 Seeding Medium	HepG2 Growth Medium:
80 mL RPMI	450 mL RPMI
20 mL FCS	50 mL FCS
0.2 mL antibiotic-antimycotic	1 mL antibiotic-antimycotic

PROCEDURE

- Thaw cryovial containing the cells in a water bath at 37°C.
- Seed $\sim 2.5 \times 10^3$ cells per cm^2 (for experiment, seed 2.5×10^4 cells per cm^2) with seeding medium.
- The next day wash the cells with PBS and change medium (growth medium).
- Change medium every second day.
- Split the cells Monday and Friday ($\sim 80\%$ confluence).
- To split the cells remove medium and wash the cells with warm PBS.
- Remove PBS and add 10 mL $1\times$ trypsin for 5−7 min at 37°C; check under the microscope.
- When the cells detach, add 10 mL PBS and transfer the mixture to a Falcon tube.
- Centrifuge for 5 min at 1.000 rpm, remove supernatant, and add 1 mL growth medium.
- Split the cells 1:4 or count the cells and seed them in 14.5 cm petri dishes for further cultivation or, e.g., in 6-well dishes for experiments.

CULTURE OF INS-1 CELLS AND ANALYSIS OF INSULIN SECRETION

Materials	Distributor/ Manufacturer	Cat. No.	Storage
RPMI 1640	Gibco	21875-091	4°C
FCS	Gibco	10270106	−20°C
Antibiotic-Antimycotic (100×)	Gibco	15240-062	−20°C
Sodium Pyruvate MEM (100 mM)	Gibco	11360-039	4°C
β-Mercaptoethanol	Sigma	M-7522	RT
RPMI 1640 w/o glucose	Gibco	11879-020	4°C
D (+) Glucose	Merck	1.08337	RT
Albumin Fraction V (BSA, fatty acid−free)	Roth	0052.3	4°C
NaCl	AppliChem	A3597	RT
KCl			RT
NaHCO$_3$			RT
NaH$_2$PO$_4$			RT
MgCl$_2$			RT
CaCl$_2$			RT
HEPES	Sigma	H3375	RT
Rat Ultrasensitive Insulin ELISA	Mercodia	10-1251-01	4°C

Preparation of Solutions and Buffers

1. Glucose stock, 1 M:
   Dissolve glucose in distilled H_2O and filter the solution using a sterile
   syringe filter 0.2 µm. Store at 4°C
2. BSA stock fatty acid–free, 5%
   Dissolve BSA in distilled H2O and filter the solution using a sterile syringe
   filter 0.2 µm. Store at 4°Cx
3. INS1 growth medium:
   500 mL RPMI
   50 mL heat-inactivated FCS
   1 mL antibiotic-antimycotic
   5 mL sodium pyruvate MEM (100 mM)
   1.57 µL β-mercaptoethanol
4. INS1 starvation medium
   500 mL RPMI w/o glucose
   2.75 mL glucose (1 M, sterile filtered)
   1 mL antibiotic-antimycotic
   5 mL sodium pyruvate MEM (100 mM)
   10 mL BSA (5%, fatty acid–free)
   1.57 µL β-mercaptoethanol
5. KRBH Buffer (stock)   for 1 L: store at 4°C

	for 1 L: store at 4°C
135 mM NaCl	7.9 g
3.6 mM KCl	0.268 g
5 mM $NaHCO_3$	0.42 g
0.5 mM NaH_2PO_4	0.069 g
0.5 mM $MgCl_2$	0.22 g
1.5 mM $CaCl_2$	0.22 g
10 mM HEPES	1 mL of 1 M stock solution

   Before each experiment add 0.1% BSA (fatty acid–free) to the required volume of
   KRBH buffer.

Procedure

Seed cells in 6-well dishes (3×10^5 cells/well) and culture them for 3 days in
growth medium until they reach $\sim$ 60% confluence.
Change to starvation medium overnight.
Split the cells Monday and Friday ($\sim$80% confluence).
Wash the cells twice with KRBH buffer (1 mL/well).
Incubate the cells for 1 h in KRBH buffer (1 mL/well).
Stimulate the cells for 30 min with different concentrations of glucose (e.g., 0, 2,
5, 10, 15, 20 mM).

Collect the supernatant, centrifuge it for 3 min at 3000 rpm, and store it at $-20°C$ until analysis.

Lyse the cells to determine protein concentration/well.

Measure insulin concentration in the supernatants by ELISA using a dilution of 1:100–1:120.

ELECTRICAL STIMULATION OF MUSCLE CELLS

Electrical Pulse Stimulation of Human Skeletal Muscle Cells

Consumables, buffer, media, and solutions

Materials	Distributor/Manufacturer	Cat. No.	Storage
Ham's F-12 Nutrient Mix, powder	Gibco	21700-026	4°C
Alpha MEM, powder	Gibco	11900-016	4°C
FCS	Gibco	10270106	$-20°C$
Supplement Pack for skeletal muscle cells	PromoCell	C-39360	$-20°C$
Antibiotic-Antimycotic (100×)	Gibco	15240-062	$-20°C$
Horse serum	Gibco	16050-122	$-20°C$
PBS, no calcium, no magnesium	Gibco	14190-169	4°C or RT
C-Pace EP Cell Culture Stimulator	IonOptix		
C-Dish	IonOptix		

Preparation of Solutions and Buffers

1. Sterilization of C-Dishes:

   After every use, rinse the C-Dishes with distilled H_2O and let them dry. Put them in a beaker and heat them for 4 h at 100°C.

   After every fifth use, soak the C-Dishes in distilled H_2O overnight, let them dry, and proceed as described above.

2. hSkMC growth medium:
   - 500 mL F12/Alpha MEM stock
   - 1 pack of "Supplement Pack for skeletal muscle cells"
   - 1 mL antibiotic-antimycotic

3. hSkMC Diff (+) medium:
   - 500 mL Alpha MEM stock
   - 10 mL Horse serum
   - 1 mL antibiotic-antimycotic

4. hSkMC Diff (−) medium:
   - 500 mL Alpha MEM stock
   - 1 mL antibiotic-antimycotic

Procedure

- Seed the cells in 6-well dishes using growth medium; change medium every second day.
- Allow them to reach 90%–100% confluence, then switch to Diff (+) medium.
- Wash the cells 1× with warm PBS at day 5, add Diff (−) medium overnight (this step should be done late in the afternoon to shorten starvation time. Long starvation time leads to, e.g., less effect on AMPK phosphorylation and myokine secretion.)
- Add fresh Diff (−) medium on day 6 before starting the EPS protocol.
- Add C-Dishes to the 6-well plate.
- Connect C-Dish electrodes with C-Pace EP Cell Culture Stimulator via 9-connector ribbon cables. The cable is thin enough to close the incubator door on and maintain a good seal.
- Settings for C-Pace EP Cell Culture Stimulator:
  - 2 ms duration
  - 11.5 V
  - 1 Hz
- To start the stimulation, switch from Disabled to Enable in the menu.

Electrical Pulse Stimulation of C2C12 Muscle Cells
Consumables, buffer, media, and solutions

Materials	Distributor/ Manufacturer	Cat. No.	Storage
DMEM (high glucose, pyruvate)	Gibco	41966	4°C
FCS	Gibco	10270106	−20°C
Antibiotic-Antimycotic (100×)	Gibco	15240-062	−20°C
Horse serum	Gibco	16050-122	−20°C
PBS, no calcium, no magnesium	Gibco	14190-169	4°C or RT

PREPARATION OF SOLUTIONS AND BUFFERS

1. Sterilization of C-Dishes:

   After every use, rinse the C-Dishes with distilled H_2O and let them dry. Put them in a beaker and heat them for 4 h at 100°C.

   After every fifth use, soak the C-Dishes in distilled H_2O overnight, let them dry and proceed as described above.
2. C2C12 growth medium:
   500 mL DMEM
   50 mL FCS
   1 mL antibiotic-antimycotic

3. C2C12 Diff (+) medium:
   500 mL DMEM
   10 mL Horse serum
   1 mL antibiotic-antimycotic
4. C2C12 Diff (−) starvation medium:
   500 mL DMEM
   1 mL antibiotic-antimycotic

Procedure

1. Seed the cells in 6-well dishes using growth medium; change medium every second day.
2. Allow them to reach 80% confluence, then switch to Diff (+) medium.
3. For EPS protocol by Lambernd et al., wash the cells 1× with warm PBS at day 5, add Diff (−) medium overnight, and give fresh Diff (−) medium the next day before starting the EPS protocol.
4. For EPS protocol by Burch et al., add fresh Diff (+) medium before starting the EPS protocol at day **6**.
5. Start the EPS.

EPS Protocols for C2C12

1. EPS by Lambernd et al.: 4−24 h, 1 Hz, 2 ms, 11.5 V.
2. EPS by Burch et al.: 24 h, 1 s (50 Hz, 1 ms, 14 V), 1 s (break).

Rat Cardiomyocyte Contraction
Materials
Consumables, buffer, media, and solutions

Materials	Distributor/Manufacturer	Cat. No.	Storage
μ-Dish 35 mm, high	ibidi GmbH	81156	RT
Laminin	Roche Diagnostics—Applied Science	11243217001	−20°C
Fura-2AM	Calbiochem/Merck Millipore	344905	−20°C
Ascorbic Acid	Sigma	A4544	RT
DL-Isoproterenol hydrochloride	Sigma	I5627	RT
DMEM F12			
FCS			
Pantothenate			
Insulin	Sigma	I5523	

Preparation of Solutions and Buffers
Animals

Lewis rats 6−12 weeks old (150−350 g/animal)

Fura Working Solution

Stock solution:

Dissolve 100 μg Fura-2AM powder in 1 mL DMSO.
Stock concentration: 100 μg/mL.
Working solution:
1:1000 dilution of Fura-2AM stock solution with adipocyte medium.
Final concentration: 100 ng/mL.

Laminin

Stock solution:

0.5 mSg/mL
Working solution:
Use 40 μL stock solution in 1 mL.
Coat the ibidi dish with 800 μL/dish.
Laminin concentration of the working solution: $2-5$ μg/cm^2.

Adipocyte Medium

DMEM/F12
33 μmol/L biotin
17 μmol/L pantothenate

Modified Tyrode's Solution

	Final Concentration	g/L	Art. No.	Supplier
NaCl	125.0 mM	7.305	39571.	Carl Roth GmbH
KH$_2$PO$_4$	1.2 mM	0.163	4871	Merck
KCl	2.6 mM	0.194	4936	Merck
MgSO$_4$*7H$_2$O	1.2 mM	0.296	5833	Merck
CaCl$_2$*2H$_2$O	1.0 mM	0.147	2382	Merck
Glucose	10.0 mM	1.982	1.08342	Merck
HEPES	10.0 mM	2.383	242608	Roche

pH = 7.4

Isoprenaline

Stock solution:

Prepare a 10 mM stock solution of ascorbic acid in bidest water (dissolve 0.0176 g ascorbic acid in 10 mL water). Prepare a 10 mM stock solution of DL-isoproterenol (dissolve 0.02477 g in 10 mL water). Mix both solutions in a ratio of 1:1. Dilute 1:5 with bidest water to a concentration of 1 mM. Prepare fresh for each experiment.

Working solution:
Dilute 1 mM stock solution 1:1000 (e.g., 1 μL stock in 999 μL medium). This results in a 1 μM isoprenaline solution.
Dilute the 1 μM isoprenaline solution 1:100 to achieve a final concentration of 10 nM (e.g., use 10 μL of the 1 μM isoprenaline solution on 990 μL Tyrode's solution in the ibidi dishes).

Insulin

(Preparation of insulin stock solution is described in separate SOP)
Insulin stock solution:
100 μM
Insulin working solution:
Dilute stock solution 1:10 (e.g., 20 μL stock in 180 μL medium).
Use 10 μL working solution in 1 mL medium.
Results in a final concentration of 100 nM.

Procedure

Isolation and seeding of rat cardiomyocytes:
- Cardiomyocytes are isolated from Lewis rats (Lew/Crl) as described in Eckel et al. (Biochem J 1983).
- Cells were seeded at a density of about 100,000 cells/dish (70,000–130,000 cells/dish) on laminin-coated dishes and cultured overnight.

Loading of cardiomyocytes with Fura-2AM:
- Prepare 2 mL working solution of Fura-2AM (dissolve 2 μL stock solution in 2 mL prewarmed medium).
- Discard medium from the dish.
- Add 2 mL Fura-2AM working solution to the cells.
- Incubate for 25 min at RT in the dark (blackened dish).
  (Fura-2AM is a pentaacetoxymethyl ester; the protection of carboxylic groups of Fura as acetoxymethyl [AM] esters makes the dye neutral, so it can cross the cell membrane. Once inside the cell, esterases will cleave AM groups.)
- Use for all following washing steps prewarmed medium (37°C).
- Washing of the cells to wash out Fura-2AM that has not been taken up by the cells:
  - 2 times shortly with 1 mL medium
  - 1 time 5 min with 2 mL medium at RT in the dark
- Discard medium.
- Add 1 mL medium and incubate for 20 min at RT in the dark.

(Complete hydrolysis of the AM esters leads to free dye and is very important to avoid artefacts. If the experiment begins before all AM ester dye is converted to free dye, total concentration increases during the experiment and gives place to false fluorescence variations.)
- Start measurement.

Cell treatment:

- When Fura-2AM treatment is skipped, change the medium from M199 to modified Tyrode's solution directly before the treatment.
- The cells are directly treated with isoprenaline before measurement.

Measurement of contraction:

Italic points are for measurement with Fura-2AM.

- Start the computer and IonWizard.
- Switch on the heating plate early enough that the plate can warm up already (37°C).
- Switch on all devices from the right side to the left side!
- Transfer the dish to the microscope on the heating plate.
- Put both electrodes in the medium.
- Start contraction of the cells with 1 Hz, 0.5 ms, and 15 V.
- Start measurement.
- Look for well-contracting cardiomyocytes with rod shape.
- *Navigate the cell in the field of analysis and close the shutters so that just the whole cardiomyocyte is visible.*
- Press "Start" in the program and record the contraction (sarcomere *length and Fura ratio*).
- Trace → Automatic Limited → Off → On (defines a range automatically).
- Measure at least 10 different cardiomyocytes per dish.
- Measure at least seven contractions per cardiomyocyte and take the last five contractions for analysis if possible.
- *When Fura-2AM is applied to the cells, always measure the background close to the cell after measuring the contraction of the cardiomyocyte.*
- After measuring at least 10 different cardiomyocytes, save the file (File, Save as, Experiment No., and Treatment).

MEASUREMENT OF OXYGEN CONSUMPTION (ADIPOCYTES) USING THE SEAHORSE SYSTEM

Materials

Consumables, buffer, media, and solutions

Materials	Distributor/ Manufacturer	Cat. No.	Storage
DMEM powder	Sigma	D5030	4°C
L-Glutamine	Sigma	G-8540	RT
Sodium Pyruvate	AppliChem	A4859	RT
D-Glucose	AppliChem	A0883	RT
XFe96 FluxPak (contains 18 sensor cartridges + cell culture plates)	Seahorse	102416-100	RT
XF96 V3 PS Cell Culture Microplates (order in case the plates delivered with the FluxPak are empty)	Seahorse	101085-004	RT

Preparation of Solutions and Buffers

DMEM	Dissolve one bottle of D5030 powder in 990 mL millipore water
	Adjust pH to 7.4 ± 0.05 at 37°C
	Sterile filter and store at 4°C
	Alternatively XF-base medium can be used (102353-100)
45% D-Glucose	45 g D-Glucose
	100 mL water
	Water has to be heated before adding the glucose
	Store solution at 4°C
200 mM Glutamine:	2.92 g L-Glutamine
	100 mL water
	Prepare aliquots and store at −20°C
200 mM sodium pyruvate:	2.2 g sodium pyruvate
	100 mL water
	Prepare aliquots and store at −20°C

Procedure

I. Seeding of Cells

- Human preadipocytes are seeded at a density of 10,000 cells/well in a seahorse XF96 cell culture microplate (the area per well is smaller compared with usual 96-well plates). Cells can be started for differentiation 2−3 days after seeding. Differentiated adipocytes should be measured at day 12−13 of differentiation (it is better to measure them a bit earlier to avoid cells that detach in these small wells).
- All wells in the corner have to stay cell-free! These wells serve as background!

NOTE:

The cell density has to be tested for each cell type. If fewer preadipocytes are seeded the respiration per well is too low to properly monitor oxygen consumption. Other cell types, which do not differentiate and can be seeded 1−2 days before the assay, are usually seeded at densities between 30,000 and 50,000.

II. Design of the Assay/Preparing the Template for the Run

- Information feeding the assay design file contain the following:
  - Cell type
  - Treatment
  - Composition of assay medium
  - Injection strategy (which ports contain which compound at which concentration)
  - Measurement protocol (here the ports used for injection have to be confirmed, mixing and measurement times are set, and number of cycles is set)

- The assay design is very important; when it does not contain all information and wells are not assigned clearly, it will be hard to analyze the data (a huge excel sheet with data for each well and time point) afterward.

III. To Do the Day before the Assay

1. Prepare sensor cartridge for the XF96 assay.

- Add 200 μL of Seahorse XF calibrant (pH 7.4) (calibrant is delivered with the FluxPak) to each well of a Seahorse 96-well utility plate.
- Place sensor cartridge on top of plate and put in the small incubator at 37°C without CO_2 overnight.

2. Turn on the instrument and start XF software to allow the instrument to stabilize at 37°C.

IV. To Do on the Assay Day

1. Prepare assay medium → add compounds as follows to the XF-base medium:

- Assay medium is based on XF-base medium or DMEM D5030 from Sigma (minimal DMEM) → without bicarbonate, glucose, pyruvate, and glutamine!

Compound	Final Concentration	Stock	For 100 mL, Add From Stock
L-Glutamine	2 mM	200 mM	1 mL
Sodium pyruvate	2 mM[a]	200 mM	1 mL
Glucose	11 mM	45% (w/v)	440 μL
	17.5 mM[b]		700 μL
	25 mM		1000 μL

[a] Sodium pyruvate concentration can be varied between 0 and 2 mM (Bordicchia et al. used 1 mM for adipocytes).
[b] Glucose concentration in the assay medium should be similar to glucose concentration used for culture of a certain cell type.

- Warm up prepared medium to 37°C and adjust pH to 7.4.
- 100 mL assay medium is sufficient for one full assay plate.

2. Replace cell culture medium with assay medium.

- Wash cell 2× with assay medium and add a final volume of **180 μL** in the last step (this step is important to get rid of the bicarbonate from your normal cell culture medium).

NOTE: Although your experiment may not need the full 96-well plate, it is obligatory to fill also the cell-free wells with 180 μL medium.

3. Place cell plate for 1 h in the incubator without CO_2 before the assay.

4a. Use this 1 h incubation time to prepare mitochondrial inhibitors and load cartridge with these reagents (oligomycine, etc.).

- Add different volumes of reagents to the different ports:
Port A: 20 μL (180 μL in well)
Port B: 22.5 μL (200 μL in well)
Port C: 25 μL (222.5 μL in well)
Port D: 27.5 μL (275.5 μL in well)

- The concentration of substances in the ports is $10\times$ higher than the final concentration (after injection).
- Here is an example how to calculate and prepare the inhibitors—depending on your injection strategy:
  NOTE:
- The FCCP concentration needs to be titrated for each cell type. The above used concentration has been tested for adipocytes.
- Although your experiment may not need the full 96-well plate, it is obligatory to fill the ports for the wells without cells with medium (the injection is based on air pressure; when you use port A and B, all ports A and B of the plate have to be loaded; the nonused port C and D can stay empty over the whole plate).

4b. Calibrate the sensors before the assay.
- Open assay template (can be carried out before on your own computer).
- Load hydrated and loaded cartridge plate into the instrument
  - with the barcode facing the back and
  - the blue notch to the front and right side.
- Calibration will take approx. 30 min.

5. Start the assay after 1 h without CO_2 and calibration.
- When calibration is carried out, continue only when the plate is ready!
- Follow instructions after calibration is finished.
  - Once the door of the instrument opens, calibration plate can be removed $\rightarrow$ note: the cartridge remains inside of the instrument!
  - Quickly replace calibration plate with cell plate $\rightarrow$ the blue notch should also be in the front and right side!

6. Prepare plate for normalization after the assay (using Cyquant).
- Remove plate from the machine and place it on ice.
- Wash $2\times$ with PBS.
- Carefully dry plate by tapping it on a paper towel (how careful you are is dependent on the adherence of your cells).
- Freeze cells for at least 2 days at $-80°C$ (the freezing step is necessary for proper lysis using the Cyquant kit; see Cyquant data sheet).

NOTE: Respiration can also be normalized to protein content per well; however, this may lead to different results compared with normalization to cell number (Cyquant). Seahorse Biosciences provides information regarding different methods to normalize.

V. Data Analysis

1. Normalization using Cyquant:
- The method is based on a fluorescent dye. Work in the dark! Or protect reagents from light.
- Prepare Cyquant lysis buffer according to the manufacturer's protocol.
- Defrost assay plate and add 200 µL of lysis buffer per well; incubate for 2—5 min on a shaker.

- Transfer 190 μL of lysis buffer from each well to the corresponding well of a black 96-well plate.
- Measure fluorescence at the Tecan reader: Exc 480 nM, Em 520 nM.
- Interpolate the intensities to cell number using a standard curve you have prepared before (see Cyquant data sheet).
  NOTE: Prepare a new curve for each new kit you use. For the adipocytes, the standard curve has been performed with one preadipocyte donor, as you cannot count differentiated adipocytes.

2. Calculation of mitochondrial respiration:

- Export data from the wave software (add view → data → rate).

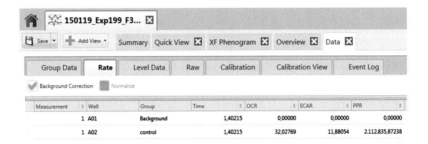

- Copy the whole table into an excel sheet.
- Normalize the OCR in each well to the cell number in this well (or mg protein).
- Calculate the components of respiration according to the Seahorse manual.

METABOLIC MEASUREMENTS

Fatty Acid Oxidation in Myotubes

Consumables, buffer, media, and solutions

Materials	Distributor/ Manufacturer	Cat. No.	Storage
AquaSafe 300	Zinsser Analytic	1008300 (5L)	RT
^{14}C Oleic Acid	Perkin Elmer	NEC317250UC	−20°C
^{14}C Palmitic Acid	Perkin Elmer	NEC075H250UC	−20°C
BSA, fatty acid—free	Sigma	A8806-5g	4°C
L-Carnitin	AppliChem	A2141	
1M NaOH	AppliChem	A1432,1000	RT
1M HCl	Carl Roth	K025.1	RT

Preparation of Solutions and Buffers

AquaSafe 300

The solution is quite viscous. Use 500 mL brown bottle and dispenser to aliquot 8 mL in scintillation tubes.

^{14}C Oleic Acid and ^{14}C Palmitic Acid

Stock: 0.1 mCi/mL in ethanol in a silanized vial.

- Dilute in human skeletal muscle differentiation medium without horse serum.
- Final concentration in the well should be 0.3 µCi/well (5.22 µM).

BSA Solution

- Stock: 2.4 mM.
- Dissolve 161 mg fatty acid–free BSA in 1 mL medium.
- Final ratio between BSA: FA = 2.5: 1.

L-Carnitine

- Stock: 1 mM.
- Dissolve 0.3954 mg L-carnitine in 2 mL of PBS.
- Store stock at −20°C.
- Final concentration in the well should be 1 µM.

Working Solution of ^{14}C Oleic Acid

Stock	^{14}C Oleic Acid Working Solution (µL)
^{14}C Fatty acid	3
BSA	1.16
L-Carnitine	0.45
Human skeletal muscle differentiation medium	44.39
Total volume	= 50

Add 50 µL working solution of ^{14}C fatty acid to 400 µL medium/well.

Procedure

hSkMCs are seeded in 48-well plates in every second row. This is important for the use of the oxidation chamber. Seed cells with a density of 20,000–30,000 cells/well. Cells are differentiated and treated for indicated time points. Use for each condition four wells to generate replicates.

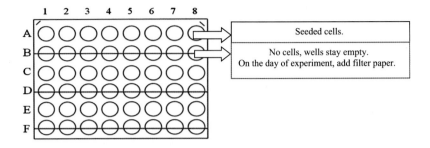

Preparations in the Laboratory

- Prepare your working place with enough cellulose papers.
- Prepare scintillation vials; add 10 mL AquaSafe 300 with the dispenser per vial.
- Label scintillation vials.
- Prepare working solution of ^{14}C fatty acid.

Assay Procedure

Day 1

- Remove medium, wash once with fresh medium, and add 400 µL human skeletal muscle differentiation medium (without horse serum) to wells that contain cells.
- Add one filter paper per free well.
- Add 50 µL 1M NaOH to each filter paper.
- Add 50 µL ^{14}C fatty acid to each well.
- Add plates without lids to the oxidation chamber (when fewer than four plates are used, add free plates to the chamber to have a final number of four plates in the chamber), add silicon sealing, put the oxidation chamber lid on top, and fix the chamber by tightening the screws.
- The last two steps should be quite quick to avoid release of $^{14}CO_2$.
- Control samples are directly stopped by the addition of 400 µL 1 M HCl.
- Fill syringe with 1M HCl, add cannula, eliminate air, puncture silicon sealing, and add 400 µL HCl to the control wells.
- Attention: HCl might drop out of the cannula. Have some paper towels to catch the HCl drops. Drop might contain radioactivity and would contaminate the chamber.
- Put the oxidation chamber for 4 h at 37°C 5% CO_2.
- Add 400 µL 1M HCl to each well containing cells; this leads to the release of $^{14}CO_2$ that is trapped on the NaOH-soaked filter papers.
- Keep oxidation chamber closed and put it overnight in the incubator.

Day 2

- Open oxidation chamber and transfer filter papers to scintillation vials.
- Add 50 µL 14C fatty acid working solution to scintillation vials as positive controls.

- Srew the lid on the vial.
- Agitate thoroughly.
- Before starting the counter, leave the vials for 2—3 h (otherwise, the background will be too high).

Cleaning of the Laboratory

- Collect all radioactive supernatants in 50 mL Falcon tubes and discard the Falcons in the proposed bins.
- Collect all radioactive plasticware and discard it in the proposed bins.

Measuring Radioactivity

- Count the radioactivity for 30 min per vial (shorter counting will lead to higher standard deviations).

Glucose Oxidation in Myotubes

Materials	Distributor/ Manufacturer	Cat. No.	Storage
AquaSafe 300	Zinsser Analytic	1008300 (5L)	RT
^{14}C glucose, uniformly labeled	Perkin Elmer	NEC042X050UC	4°C
DMEM, low glucose	Invitrogen	22320-030	4°C
1M NaOH	AppliChem	A1432,1000	RT
1M HCl	Carl Roth	K025.1	RT

Preparation of Solutions and Buffers
AquaSafe 300
The solution is quite viscous. Use 500 mL brown bottle and dispenser to aliquot 8 mL in scintillation tubes.

^{14}C Glucose, Uniformly Labeled
Stock: 310 mCi/mL in ethanol:H_2O (9:1) in a silanized vial.

- Dilute in DMEM, low glucose.
- 51.46 μL ^{14}C glucose in 1 mL DMEM, low glucose.
- 360 μL working solution per well.
- Final concentration in the well should be 0.2 μCi/well.

Insulin Working Solution

- Separate SOP for preparing stock solution.
- Stock 100 μM, storage −20°C.
- Dilute stock 1:10; add 20 μL of 1:10 dilution per 2 mL medium.
- 360 μL/well.

Procedure

hSkMCs are seeded in 48-well plates in every second row. This is important for the use of the oxidation chamber. Seed cells with a density of 20,000−30,000 cells/well. Cells are differentiated and treated for indicated time points. Use for each condition four wells to generate replicates.

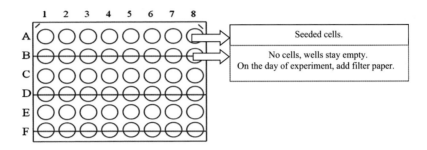

Preparations in the Laboratory

- Prepare your working place with enough cellulose papers.
- Prepare scintillation vials; add 8 mL AquaSafe 300 with the dispenser per vial.
- Label scintillation vials.
- Prepare working solution of ^{14}C glucose.

Assay Procedure
Day 1

- Pemove medium, wash one with fresh medium, and add 360 µL DMEM (low glucose!) to wells that contain cells.
- Prepare insulin working solution for stimulation and add 360 µL DMEM + insulin to wells.
- Assay has to be down in low-glucose medium, otherwise cells use the nonlabeled glucose for oxidation.
- Add one filter paper per free well.
- Add 50 µL 1M NaOH to each filter paper.
- Add 40 µL ^{14}C glucose working solution to each well.
- Add plates without lids to the oxidation chamber (when less than four plates are used, add free plates to the chamber to have a final number of four plates in the chamber), add silicon sealing, put the oxidation chamber lid on top, and fix the chamber by tightening the screws.
- The last two steps should be quite quick to avoid release of ^{14}CO$_2$.
- Control samples are directly stopped by the addition of 400 µL 1 M HCl.
- Fill syringe with 1M HCl, add cannula, eliminate air, puncture silicon sealing, and add 400 µL HCl to the control wells.

- Attention: HCl might drop out of the cannula. Have some paper towels near the wells to catch the HCl drops. Drop might contain radioactivity and would contaminate chamber.
- Put the oxidation chamber for 4 h at 37°C 5% CO_2.
- Add 400 µL 1M HCl to each well containing cells.
- These lead to the release of $^{14}CO_2$, which is trapped on the NaOH-soaked filter papers.
- Keep oxidation chamber closed and put it overnight in the incubator.

Day 2

- Open oxidation chamber and transfer filter papers to scintillation vials.
- Add 50 µL 14C glucose working solution to scintillation vials as positive controls.
- Srew the lid on the vial.
- Agitate thoroughly.
- Before starting the counter, leave the vials for 2−3 h (otherwise, the background will be too high).

Cleaning of the Laboratory

- Collect all radioactive supernatants in 50 mL Falcon tubes and discard the Falcons in the proposed bins.
- Collect all radioactive plasticware and discard it in the proposed bins.

Measuring Radioactivity

- Count the radioactivity for 30 min per vial (shorter counting will lead to higher standard deviations).

Glucose Oxidation in Myotubes After Pulse Stimulation

Consumables, buffer, media, and solutions

Materials	Distributor/ Manufacturer	Cat. No.	Storage
AquaSafe 300	Zinsser Analytic	1008300 (5L)	RT
^{14}C glucose, uniformly labeled	Perkin Elmer	NEC042X050UC	4°C
DMEM, low glucose	Invitrogen	22320-030	4°C
1M NaOH	AppliChem	A1432,1000	RT
1M HCl	Carl Roth	K025.1	RT

Preparation of Solutions and Buffers
AquaSafe 300

The solution is quite viscous. Use 500 mL brown bottle and dispenser to aliquot 8 mL in scintillation tubes.

^{14}C Glucose, Uniformly Labeled

Stock: 310 mCi/mL in ethanol:H_2O (9:1) in a silanized vial.

- Dilute in DMEM, low glucose.
- 51.46 μL ^{14}C glucose in 1 mL DMEM, low glucose.
- 360 μL working solution per well.
- Final concentration in the well should be 0.2 μCi/well.

Insulin Working Solution

- Separate SOP for preparing stock solution.
- Stock 100 μM, storage −20°C.
- Dilute stock 1:10; add 20 μL of 1:10 dilution per 2 mL medium.
- 360 μL/well.

Procedure

Seeding of the cells:

- Primary human myoblasts are seeded in 6-well plates (one cryovial containing 2.4×10^6 cells on four 6-well plates).
- Add four coverslips per well.
- Thaw cryovial and dilute cells in 4 mL growth medium (prepare the cell suspension in a 50-mL Falcon tube).
- Add 50 μL cell suspension per coverslip.
- Leave under the hood for 10−15 min (cells attach rapidly to the coverslips).
- Add to the remaining cell suspension 48 mL growth medium.
- Add 2 mL cell suspension per well.
- Allow them to reach 90%−100% confluence, then switch to differentiation medium containing 2% horse serum for 5 days.
- Change to differentiation medium without horse serum on day 5 of differentiation in the morning.
- Start EPS in the evening (1 Hz, 2 ms, 11.5 V).
- Stop EPS on day 6 in the morning.
- Transfer coverslips to 48-well plate.

Day 1

Use for each condition cells of four coverslips to generate replicates.

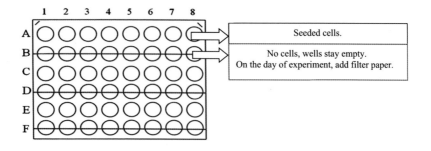

Preparations in the Laboratory

- Prepare your working place with enough cellulose papers.
- Prepare scintillation vials; add 8 mL AquaSafe 300 with the dispenser per vial.
- Label scintillation vials.
- Prepare working solution of ^{14}C glucose.
- Assay procedure.

Assay Procedure

Day 1

- Remove medium, wash one with fresh medium, and add 360 μL DMEM (low glucose!) to wells that contain cells.
- Prepare insulin working solution for stimulation and add 360 μL DMEM + insulin to wells.
- Assay has to be down in low-glucose medium, otherwise cells use the nonlabeled glucose for oxidation.
- Add one filter paper per free well.
- Add 50 μL 1M NaOH to each filter paper.
- Add 40 μL ^{14}C glucose working solution to each well.
- Add plates without lids to the oxidation chamber (when less than four plates are used, add free plates to the chamber to have a final number of four plates in the chamber), add silicon sealing, put the oxidation chamber lid on top, and fix the chamber by tightening the screws.
- The last two steps should be quite quick to avoid release of $^{14}CO_2$.
- Control samples are directly stopped by the addition of 400 μL 1 M HCl.
- Fill syringe with 1M HCl, add cannula, eliminate air, puncture silicon sealing, and add 400 μL HCl to the control wells.
- Attention: HCl might drop out of the cannula. Have some paper towels near the wells to catch the HCl drops. Drop might contain radioactivity and would contaminate chamber.
- Put the oxidation chamber for 4 h at 37°C 5% CO_2.
- Add 400 μL 1M HCl to each well containing cells.
- These lead to the release of $^{14}CO_2$ that is trapped on the NaOH-soaked filter papers.
- Keep oxidation chamber closed and put it overnight in the incubator.

Day 2

- Open oxidation chamber and transfer filter papers to scintillation vials.
- Add 50 μL 14C fatty acid working solution to scintillation vials as positive controls.
- Srew the lid on the vial.
- Agitate thoroughly.
- Before starting the counter, leave the vials for 2−3 h (otherwise, the background will be too high).

Cleaning of the Laboratory

- Collect all radioactive supernatants in 50 mL Falcon tubes and discard the Falcons in the proposed bins.
- Collect all radioactive plasticware and discard it in the proposed bins.

Measuring Radioactivity

- Count the radioactivity for 30 min per vial (shorter counting will lead to higher standard deviations).

Glucose Uptake in Myotubes

Materials	Distributor/ Manufacturer	Cat. No.	Storage
AquaSafe 300	Zinsser Analytic	1008300 (5L)	RT
Cytochalasin B	AppliChem	A7657,0005	−20°C/dissolved at 4°C
D-^{14}C-Glucose 250 μCi (9.25 MBq)	Perkin Elmer	NEC495	−20°C
L-^{14}C-Glucose 050UC (1.85 MBq)	Perkin Elmer	NEC478	−20°C
NaOH, 1 mol/L (1 N)	VWR		RT
Acetic Acid			RT

Preparation of Solutions and Buffers
AquaSafe 300

The solution is quite viscous. Use 500 mL brown bottle and dispenser to aliquot 10 mL in scintillation tubes.

D-^{14}C-Glucose and L-^{14}C-Glucose

Stock: 100 μCi/mL (0.1 μCi/μL).

- Prepare a 1:20 dilution of the stocks in hSkMC Diff (−) medium.
- 50 μL of this dilution is required for each well.
  For example, for 5 plates with 6 wells +3 wells extra = 33 wells
  33 wells × 50 μL/well = 1.65 mL
  1:20 dilution of the stock: 82.5 μL D-^{14}C-Glucose + 1.568 mL medium
- Final concentration in the well should be 0.25 μCi/mL.
- Document the amount of radioactivity removed from the stocks in the appropriate list.

Cytochalasin B **(Very Toxic)**

Open the original bottle with tweezers.

- Dissolve 5 mg cytochalasin B in 417 μL ethanol; vortex thoroughly.
- Take 83.5 mL PBS from a fresh 500 mL bottle PBS.
- Dilute dissolved cytochalasin B in 416.5 mL PBS.
- Transfer in 500 mL bottle with dispenser.
- Final concentration of cytochalasin B = 0.25 μM.
- Stable for 2−3 weeks at 4°C.

Insulin Working Solution

- Separate SOP for preparing stock solution.
- Stock 100 μM, storage −20°C.
- Dilute stock 1:10; add 10 μL of 1:10 dilution per 1 mL medium.

Procedure

hSkMCs are seeded in 6-well plates, differentiated, and treated for indicated time points. Use for each condition three wells to generate triplicates.

Preparations in the Laboratory

- Prepare your working place with enough cellulose papers.
- Label scintillation vials for experiments +4 vials for positive + negative control.
- Fill vials with 10 mL AquaSafe 300 each using the dispenser.
- Prepare working solution of D-^{14}C-Glucose and L-^{14}C-Glucose.

Assay Procedure

- Add insulin.
- Incubate the plate for 30 min at 37°C and 5% CO_2 (for easy handling during the complete assay, plan a 3 min break between each plate).
- After 30 min, add 50 μL D-^{14}C-Glucose and L-^{14}C-Glucose working solution.
- Incubate the plate for 2 h at 37°C and 5% CO_2.
- Remove the medium and put the plate on ice.
- Wash twice with 1 mL cold cytochalasin B while on ice.
- Add 1 mL 1 M NaOH/well.
- Incubate for 30 min at 37°C and 5% CO_2.
- Add 60 μL acetic acid for neutralization; rotate plate carefully.
- Homogenize solution, rinse the well several times, and add 1 mL of the lysate in the labeled scintillation vial.
- Srew the lid on the vial.
- Agitate thoroughly.
- Before starting the counter, leave the vials for 2−3 h (otherwise, the background will be too high).

Data Analysis

Calculate the average of triplicates and subtract the background (average L-glucose) of each D-glucose average; present data as percentage of control.

Lipolysis in Human Adipocytes

Materials

Consumables, buffer, media, and solutions

Materials	Distributor/ Manufacturer	Cat. No.	Storage
DPBS 1×	Gibco	14190-094	4°C
Free Glycerol Reagent	Sigma	F6428	4°C
(−)-Isoproterenol hydrochloride	Fluka	59648	
BSA, fatty acid−free	Roth	Art. 0052.3	4°C
Sodium Pyruvate 100 mM solution	Invitrogen	11360	4°C

Preparation of Solutions and Buffers

Free Glycerol Reagent
- Dissolve in 40 mL aqua dest and store at 4°C.

Krebs Ringer HEPES (KRB) buffer

130 mM	NaCl	7.6 g/L
10 mM	HEPES	2.38 g/L
10 mM	$MgSO4 \times 7H_2O$	0.306 g/L
2.5 mM	$NaH2PO4 \times H_2O$	0.345 g/L
4.6 mM	KCl	0.343 g/L
2.5 mM	$CaCl2 \times 2H_2O$	0.368 g/L
2.5 mM	Pyruvate (100 mM)	40 mL/L

- pH 7.4
- Adjust to 1000 mL and filter sterile; keep at 4°C.

Procedure

Day 1

There are two types of treatments:

1. Treat cells with compound of interest in hSkMC Diff (−) for 24 h → effect of your compound on the lipolysis pathway.
2. Starve your cells for 24 h with SkMC Diff (−) and assess the direct effect of your compound on lipolysis by adding it to the KRB.

Day 2

- Prepare 1% BSA in KRB in a semisterile way (a sterile filter cannot be used because the filter contains glycerol, which can impair your results):
  - Pipette the appropriate volume of KRB for your experiment into a sterile Falcon tube (under the hood).
  - Weigh the BSA (unsterile).
  - Add BSA to the sterile KRB.
  - Close Falcon tube and let dissolve on the rolling shaker.
- Prepare different concentrations of noradrenaline for a dose–response curve and the treatments:
  - Noradrenaline stock: $c = 10^{-2}/10$ mM.
  - Dose–response curve from 10^{-6} bis 10^{-11}.
  - For stimulation of your treated cells, use $10^{-6}/10$ μM isoproterenol.
- Aspirate SkMC Diff (−) and add KRB ± noradrenalin; incubate for 4 h at 37°C and 5% CO_2.
  - 500 μL per well of a 12-well plate
  - 1000 μL per well of a 6-well plate

Glycerol Determination

- Prepare glycerol standard curve.

- Prepare 1 M glycerol stock: Add 72 μL of glycerol ($\sim$92.09 mg) to 100 mL aqua dest.
- Prepare at least 500 μL of the following glycerol concentrations in KRB with 1% BSA:
  - 5, 1, 0.5, 0.25, 0.1, 0.05, 0.01, 0.005 mM.
- Pipette 100 μL of glycerol standard or sample into a 96-well plate.
- Add 100 μL of free glycerol reagent per well.
- Incubate 15 min at RT in the dark.
- Read out at 540 nm.
- Interpolate values from glycerol standard curve (use XY-table in prism; analyze/ XY analysis/nonlinear/polynomial—fourth order, interpolate unknowns from standard curve).

Principle of the Method

Lipolysis: TG $\rightarrow$ FFA + Glycerol

Thus, measurement of glycerol in the supernatant reflects the rate of lipolysis.

Measurement of Reactive Oxygen Species in Cell Lysates

Consumables, buffer, media, and solutions

Materials	Distributor/ Manufacturer	Cat. No.	Storage
Medium w/o phenol red	Depending on cell type		4°C
H_2O_2	Sigma	H1009-5ML	RT
DCF	Invitrogen	C6827	−20°C, light protected
HEPES buffer			
PBS			

Preparation of Solutions and Buffers

Stock solution of DCF (dichlorofluorescein) (1 mM): Use one vial containing 50 μg and dissolve it in 8.7 μL ethanol and immediately add 80 μL of PBS.

Dilute H_2O_2 in medium to a final concentration of 0.3%.

Procedure

On the first day:

- Change cells to phenol red—free medium and add treatments as needed.

  On the second day:

- Prepare DCF and H_2O_2 solutions.
- Dilute DCF 1:100 in 1 mL medium and incubate the cells with this fresh medium for 30 min at 37°C.
- For positive controls, incubate cells with DCF and 0.3% H_2O_2 for 30 min.
- Wash with cold PBS and lyse cells in HEPES buffer (100 μL per well).

- Centrifuge lysates quickly and immediately measure fluorescence in a plate reader (duplicates of 50 μL, Ex/Em: ∼492–495/517–527 nm).
- Use the rest of the lysate for Bradford assay.

STAINING AND MICROSCOPY
OIL RED O
Materials
Consumables, buffer, media, and solutions

Materials	Distributor/ Manufacturer	Cat. No.	Storage
DPBS 1×	Gibco	14190-094	4°C
Formaldehyde solution for molecular biology, 36.5%–38% in H_2O	Sigma-Aldrich	F8775	RT
Oil Red O	Sigma-Aldrich	O0625-25G	RT
Picric acid solution, 1.3% in H_2O (saturated)	Sigma-Aldrich	P6744	RT
Filters (Whatman) qualitative filter paper, folded (prepleated grades), Grade 595 1/2 folded filters, 90 mm, 100/pk	Sigma-Aldrich	Z612928-100EA	

Preparation of Solutions and Buffers
Fixative

- 15 mL picric acid solution (71% v/v)
- 5 mL formaldehyde solution 37% (24% v/v)
- 1 mL acetic acid (5% v/v)
    - → *Prepare freshly for each staining!*
    - → *Discard used fixative in a separate waste bottle.*

Oil Red O Stock Solution
0.5% Oil Red O (w/v) in 99% isopropanol:

- 0.25 g Oil Red O
- 50 mL 99% isopropanol
    - → *Stock Solution can be stored for 1–2 weeks at RT in the dark.*

Oil Red O Working Solution

- 18 mL Oil Red O stock solution (60% v/v)
- 12 mL aqua dest (40% v/v)
    - → *Stock solution can be stored for 1–2 weeks at RT in the dark.*

Procedure
Day 1

- Wash 1× with prewarmed DPBS.
- Fix cells with the fixative overnight at RT.
- Prepare Oil Red working solution.

Day 2

- Filter Oil Red O working solution (put the folded filter in the glass filter; filtration may take 20 min).
- Wash 1× with DPBS.
- Then 3 × 5 min with DPBS.
- Wash cells shortly with aqua dest and dry plate by tapping it onto a paper towel.
- Differentiate cells 5 min with 40% isopropanol.
- Incubate 10 min with Oil Red O working solution.
- Remove Oil Red O working solution, differentiate 5 s with 40% isopropanol, and change immediately to aqua dest.

Take Pictures of the Cells, Then Continue with Elution of the Dye

- Wash with 1 mL (6-well) 40% isopropanol → before aspirating pipette 100 µL into a 96-well plate in duplicates.
- Elute with 1 mL of 100% isopropanol; move the plate to elute the whole dye → pipette into a 96-well plate.
- Read out at 500 nm → subtract the blank (100% isopropanol) from the samples.

NEUTRAL RED STAINING—CYTOTOXICITY TEST

Consumables, buffer, media, and solutions

Materials	Distributor/Manufacturer	Cat. No.	Storage
DPBS 1×	Gibco	14190-094	4°C
Neutral Red (powder)	Sigma-Aldrich	N4638	2-8°C
Acetic Acid	Merck Schuchardt		RT
Ethanol	Merck		RT

Preparation of Solutions and Buffers
Neutral Red Working Solution

- Final concentration: **50 µg/mL**.
- Weigh out 5 mg of neutral red and dissolve in 100 mL of SkMC Diff (−) medium.*
  * Or any other medium your cells were treated in.
- Prewarm mixture of SkMC Diff (−) medium and neutral red.
- Filter sterile (0.2 µm) to avoid formation of dye crystals.
- **Prepare immediately before use**.

Elution Medium
= 50% ethanol supplemented with 1% acetic acid

- 49.5 mL ethanol + 49.5 mL H_2O
- 1 mL acetic acid

Procedure

- Collect or discard supernatants of each well.
- Add **1−2 mL of neutral red working solution** to each well.
  - 6-well: 2 mL
  - 12-well: 1 mL
- Incubate cells for **3 h at 37°C** in a humidified 5% CO_2/95% air atmosphere.
- Remove neutral red medium.
- Wash cells carefully twice with 1× PBS (RT).
- Add **500 µL−2LmL elution medium** to each well to extract dye.
  - 12-well: 500 µL
  - 6-well: 1−2 mL
- Shake plates gently on orbital shaker for **10 min at RT**.
- Pipette directly ~ 200 µL into a 96-well plate.
- Read out at **540 nm**.
  - Subtract blank (elution medium) from the samples.

NILE RED STAINING

Consumables, buffer, media, and solutions

Materials	Distributor/ Manufacturer	Cat. No.	Storage
DPBS 1×	Gibco	14190-094	4°C
Formaldehyde solution for molecular biology, 36.5%−38% in H_2O	Sigma-Aldrich	F8775	RT
Picric acid solution, 1.3% in H_2O (saturated)	Sigma-Aldrich	P6744	RT
DAPI (4,6-diamino-2-phenylindole)	Sigma	D9542	−20°C
ProLong gold antifade	Life Technologies	P36930	−20°C
Gelatin from porcine skin, BioReagent, Type A, powder	Sigma	G1890-100G	RT

Nile Red stock solution (1 mg/mL): Dissolve 25 mg of Nile Red (Invitrogen N-1142) in 25 mL of DMSO; store at RT in the dark.

Preparation of Solutions and Buffers
Fixative

- 15 mL picric acid solution (71% v/v)
- 5 mL formaldehyde solution 37% (24% v/v)

- 1 mL acetic acid (5% v/v)
  - → *Prepare freshly for each staining!*
  - → *Discard used fixative in a separate waste bottle.*

DAPI Stock
5 mg/mL in deionized water

- Aliquots at $-20°C$.

Gelatine Solution for Coating of Coverslips
0.1% gelatine (w/v) in $1\times$ DPBS

- Add 0.5 g gelatine to 500 mL $1\times$ DPBS.
- Autoclave solution to dissolve and sterilize (prepare two bottles or one 1000 mL bottle → gelatin might boil over).
- Sterile solution can be stored at 4°C.

Procedure
Optionally, coat coverslips before seeding the preadipocytes.

- Put coverslips with sterile tweezers into a 12-well plate, or even better three coverslips per well of a 6-well plate.
- Incubate with 0.1% gelatine solution for at least 1 h at 37°C.
- Aspirate 0.1% gelatine solution and let dry for 1−2 min before seeding the cells.

Day 1
- Wash $1\times$ with prewarmed DPBS.
- Fix cells with the fixative for at least 2 h or overnight at RT.

Day 2
- Wash $1\times$ with DPBS.
- Then 3×5 min with DPBS.
- Incubate 20 min with 10 µg/mL Nile Red in DPBS in the dark.
- Wash 3×5 min with DPBS.
- Incubate with 1 µg/mL DAPI in DPBS for 10 min at RT.
- Wash $2\times$ with DPBS for 5 min at RT.
- Wash $1\times$ with deionized water; leave coverslips in water before mounting.
- mounting:
  - Clean microscope slides with 70% EtOH.
  - Using fine tweezers, take coverslip and drain excess fluid on a paper towel.
  - Pipette 15 µL of gold antifade mounting medium on the cleaned objective slide and place coverslip (cells down) on the droplet of mounting medium (to pipette the mounting medium, use a 20−200 µL tip that has been cut off at the edge to increase the opening; the gold antifade is frozen at $-20°C$ → take it out 10 min before embedding your cells).

- Let dry for 1 day at RT in the dark.
- Store at 4°C or −20°C for long-term storage.

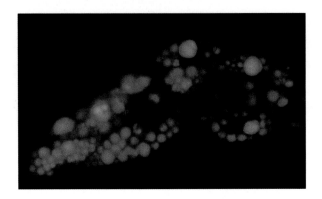

IMMUNOFLUORESCENCE STAINING OF MYOTUBES

Consumables, buffer, media, and solutions

Materials	Distributor/ Manufacturer	Cat. No.	Storage
PBS w/o Mg, Ca	Gibco	14190-169	4°C
Paraformaldehyde	AppliChem	P6148-500G	4°C
Triton X-100	Sigma	234729	RT
BSA Albumin Fraction V	AppliChem		4°C
Anti-α-Actinin	Sigma	A7732	−20°C
Goat Anti-Mouse FITC	Santa Cruz	sc-2082	4°C
Propidium Iodide 1 mg/mL	Molecular Probes	P3566	
RNase			
Vectashield	Vector Laboratories	H-1300	4°C
DAPI (4,6-diamino-2-phenylindole)	Sigma	D9542	−20°C
ProLong gold antifade	Life Technologies	P36930	−20°C

Preparation of Solutions and Buffers
Propidium Iodide

Stock   1 mg/mL   1.5 mM
Working solution   1 μM

4% PFA Solution

- Add 2 g paraformaldehyde to 50 mL PBS in a glass flask and cover the flask.
- Heat to 80°C in a hood (takes several hours).
- (Paraformaldehyde is the polymerization product of formaldehyde; heating breaks down paraformaldehyde to formaldehyde.)
- Dilute 1:2 before use.
- Final concentration should be 2% PFA.

0.2% Triton X-100

- Add 100 µL Triton X-100 to 50 mL PBS.

1% BSA Solution

- Dissolve 0.5 g BSA in 50 mL PBS.

20× SSC Buffer

- 3M NaCl.
- 0.3 M sodium citrate.
- Dilute 1:20 before use.

Procedure

Cells are normally grown on coverslips in 6 wells. If not otherwise stated, use 1 mL per step.

- Remove medium.
- Wash twice with PBS (37°C).
- Dilute 3% PFA solution.
- Add 3% PFA solution to each well for 15 min at RT.
- Wash 3× with PBS for 5 min at RT.
- Add 0.2% Triton X-100 working solution for 10 min at RT.
- Wash 3× with PBS for 5 min at RT.
- Block with 1% BSA working solution for 30–60 min at RT.
- Dilute primary antibody (test 1:50–1:100) in 1% BSA working solution.
- Incubate in well (~300 µL for 12-well plate) or carry out the following:
- *Dilute α-actinin antibody 1:100 in 1% BSA working solution.*
- *Pipette 100 µL on parafilm (100 µl should form a drop).*
- *Get the coverslip out of the well and transfer it on top of the drop (side of the coverslip with the attached cells on the parafilm); incubate for 45 min.*
- Incubate for 60 min at RT or overnight at 4°C.
- Wash 3× with PBS for 5 min at RT.

- Dilute secondary antibody in 1% BSA working solution and incubate for 1 h at RT **in the dark** in the well or carry out the following:
- Dilute goat antimouse FITC antibody 1:300 in 1% BSA working solution.
- Pipette 100 μL on parafilm (100 μL should form a drop).
- Get the coverslip out of the well and transfer it on top of the drop (side of the coverslip with the attached cells on the parafilm) and incubate for 30 min at RT.
- **Perform all following steps in the dark.**
- Wash 3× with PBS for 5 min at RT.

DAPI

- Incubate with 1 μg/mL DAPI in PBS for 5—10 min at RT.
- Wash 3× with PBS for 5 min at RT.
- Qash 2× with deionized water; leave coverslips in water before mounting.

Propidium iodide

- Wash twice with SSC buffer.
- Add 100 μg/mL RNase for 20 min at 37°C.
- Wash twice with SSC buffer.
- Dilute propidium iodide stock solution 1:1500.
- Incubate.
- Wash twice with SSC buffer.
- Mounting with Vectashield.
- Fix coverslips on microscope slides with nail polish.
- Mounting
  - Clean microscope slides with 70% EtOH.
  - Using fine tweezers, take coverslip and drain excess fluid on a paper towel.
  - Pipette 15 μL of gold antifade mounting medium on the cleaned objective slide and place coverslip (cells down) on the droplet of mounting medium (to pipette the mounting medium, use a 20—200 μL tip that has been cut off at the edge to increase the opening; the gold antifade is frozen at −20°C → take it out 10 min before embedding your cells).
- Let dry for 1 day at RT in the dark.
- Store at 4°C or −20°C for long-term storage.

Secondary Antibody	Company	Cat. No.	Dilution
Alexa Fluor 555 Goat Anti-Rabbit IgG, 0.5 mL	Life Technologies	A21428	1:500
Alexa Fluor 488 Donkey Anti-Mouse IgG (H+L) *2 mg/mL, (0.5 mL)	Life Technologies	A21202	1:500

IMMUNOFLUORESCENCE—HUMAN ADIPOCYTES

Materials

Consumables, buffer, media, and solutions

Materials	Distributor/ Manufacturer	Cat. No.	Storage
DPBS 1×	Gibco	14190-094	4°C
BSA, fatty acid–free	Roth	Art. 0052.3	4°C
Paraformaldehyde	Sigma	P6148-500G	4°C
DAPI (4,6-diamino-2-phenylindole)	Sigma	D9542	−20°C
ProLong gold antifade	Life Technologies	P36930	−20°C
Coverslips Ø 15 MM NR.1 1 * 1.000 ST	VWR	631-1579	RT
Gelatine from porcine skin, BioReagent, Type A, powder	Sigma	G1890-100G	RT

Preparation of Solutions and Buffers

3% Formaldehyde Solution (Fixative)

3% PFA (w/v) in 1× PBS

- Slowly heat 200 mL PBS to 70°C, in a fume hood.
- Add 6 g PFA and stir until the solution is dissolved and clear → can take up to 2 h.
- Filter solution.
- Make aliquots of an appropriate volume and store at −20°C.

Permeabilization and Blocking Solution

1% BSA (w/v)

0.1% Triton X-100 (v/v)

In 1× PBS, carry out the following:

- Add BSA to PBS.
- Cut the edge of the tip for Triton X-100, pipette Triton X-100 slowly, and throw pipette into the DPBS—BSA solution.
- Stir at RT until everything is dissolved.
- Has to be made freshly for each staining.

Gelatine Solution for Coating of Coverslips

0.1% gelatine (w/v) in 1× DPBS

- Add 0.5 g gelatine to 500 mL 1× DPBS.
- Autoclave solution to dissolve and sterilize (prepare two bottles or one 1000 mL bottle → gelatin might boil over).
- Sterile solution can be stored at 4°C.

DAPI Stock

5 mg/mL in deionized water

- Aliquots are stored at $-20°C$.

Procedure

Coat coverslips before seeding the preadipocytes (optionally).

- Put coverslips with sterile tweezers into a 12-well plate, or even better three coverslips per well of a 6-well plate.
- Incubate with 0.1% gelatine solution for at least 1 h at 37°C.
- Aspirate 0.1% gelatine solution and let dry for 1–2 min before seeding the cells.

Day 1

- Rinse cells in prewarmed 1× DPBS (if you use cold DPBS, your cells will start floating and will look fuzzy).
- Fix cells with 3% formaldehyde solution for 15 min at RT.
- Aspirate fixative (i.e., the formaldehyde solution has to be gathered in a separate waste).
- Wash 3× with DPBS for 5 min at RT.
- Block and permeabilize with 1% BSA, 0.1% Triton X-100 in DPBS for 1 h at RT.
- Dilute primary antibody in permeabilization and blocking solution (dilution is dependent on the antibody) and incubate for either
  - 2 h at RT
  - or at 4°C overnight.
  - ~250 μL of antibody solution is required for one well of a 12-well plate (for 50 μL if you incubate the coverslip on piece of parafilm).

Day 2

- Wash 3× with DPBS for 5 min at RT.
- Dilute secondary antibody in permeabilization and blocking solution (see table for secondary antibodies) and incubate for 1 h at RT **in the dark (and perform all following steps in the dark)**.
  - Again with ~250 μL per well of a 12-well plate.
- Wash 3× with DPBS for 5 min at RT.
- Incubate with 1 μg/mL DAPI in DPBS for 10 min at RT.
- Wash 2× with DPBS for 5 min at RT.
- Wash 1× with deionized water; leave coverslips in water before mounting.
- Mounting:
  - Clean microscope slides with 70% EtOH.
  - Using fine tweezers, take coverslip and drain excess fluid on a paper towel.

- Pipette 15 µl of gold antifade mounting medium on the cleaned objective slide and place coverslip (cells down) on the droplet of mounting medium (to pipette the mounting medium, use a 20–200 µL tip that has been cut off at the edge to increase the opening; the gold antifade is frozen at $-20°C \rightarrow$ take it out 10 min before embedding your cells).
- Let dry for 1 day at RT in the dark.
- Store at 4°C or $-20°C$ for long-term storage.

Primary Antibody	Company	Cat. No.	Dilution
Rabbit Anti-UCP1	Abcam	ab10983	1:200
Mouse Anti-MCTO2	Abcam	ab3298	1:100

Secondary Antibody	Company	Cat. No.	Dilution
Alexa Fluor 555 Goat Anti-Rabbit IgG, 0.5 mL	Life Technologies	A21428	1:500
Alexa Fluor 488 Donkey Anti-Mouse IgG (H+L) *2 mg/mL, (0.5 mL)	Life Technologies	A21202	1:500

ANALYTICAL PROCEDURES
ISOLATION OF RNA USING QIAGEN RNEASY MINI KIT

Consumables, buffer, media, and solutions

Materials	Distributor/ Manufacturer	Cat. No.	Storage
RNeasy Mini Kit (250)	Qiagen	74106	RT
Ethanol (abs.) for Molecular Biology	AppliChem	A3678,100	RT
H_2O for Molecular Biology	AppliChem	A7398,0500	RT
Cell Scraper	Sarstedt	831830	RT
β-Mercaptoethanol	Sigma	M-7522	RT
QIAshredder spin columns (250)	Qiagen	79656	RT

Preparation of Solutions and Buffers
Before using the kit for the first time, prepare RPE buffer working solution by adding 4 volumes of ethanol (96%–100%) as indicated on the bottle.

Prepare RLT lysis buffer working solution by adding 10 μL β-mercaptoethanol per mL (can be used for 4 weeks).

Procedure (Based on the Manual of Qiagen)

1. Remove the medium from the cells and wash 1× with PBS.
2. Add 150 μL/well (6-well dish) RLT lysis buffer working solution to the cells.
3. Scrape the cells and collect the samples in 1.5 mL tubes (2 mL Eppendorf tube if the QIAcube is used; please see below).
4. Homogenize the samples by pipetting the lysate into a QIAshredder spin column placed in a 2 mL collection tube and centrifuge for 2 min at full speed.
5. Add 1 volume of 70% ethanol to the homogenized lysate and mix by pipetting (do not centrifuge).
6. Transfer up to 700 μL of the sample to an RNeasy Mini spin column placed in a 2 mL collection tube and centrifuge for 15 s at $\geq$8000 $\times$g ($\geq$10,000 rpm) at RT; discard the flow-through.
7. Add 700 μL buffer RW1 to the RNeasy spin column, centrifuge for 15 s at $\geq$8000 $\times$g ($\geq$10,000 rpm), and discard the flow-through.
8. Add 500 μL buffer RPE to the RNeasy spin column, centrifuge for 15 s at $\geq$8000 $\times$g ($\geq$10,000 rpm), and discard the flow-through.
9. Add 500 μL buffer RPE to the RNeasy spin column, centrifuge for 2 min at $\geq$8000 $\times$g ($\geq$10,000 rpm), and discard the flow-through.
10. Place the RNeasy spin column in a new 2 mL collection tube and centrifuge at full speed for 1 min.
11. Place the RNeasy spin column in a new 1.5 mL collection tube, add 30−50 μL RNase-free water directly to the spin column membrane, and centrifuge for 1 min at $\geq$8000 $\times$g ($\geq$10,000 rpm) to elute the RNA.

If the **QIAcube** is used collect the lysates in 2 mL Eppendorf tubes, prepare the QIAcube:

1. Cut off the lid of the QIAshredder spin columns.
2. Load the rotor adapters with RNeasy spin columns, QIAshredder spin columns, and 1.5 mL collection tubes and place them in the QIAcube centrifuge.
3. Place the bottles containing the buffers in the QIAcube; refill if necessary.
4. Load the tips.
5. Load the 2 mL Eppendorf tube containing the lysates to the robot.
6. Adjust the elution volume if necessary.
7. Start the program: RNeasy mini—animal cells—QIAshredder.

The RNA samples should be placed on ice after isolation. Next, measure the RNA concentration using the NanoDrop. Measure in duplicates or triplicates using 1 μL of sample per measurement. Store RNA at −20°C.

ISOLATION OF RNA USING TRIPURE IN COMBINATION WITH QIAGEN RNEASY MINI KIT FOR ADIPOSE TISSUE

Consumables, buffer, media, and solutions

Materials	Distributor/ Manufacturer	Cat. No.	Storage
RNeasy Mini Kit (250)	Qiagen	74106	RT
TriPure	Roche	11667165001	4°C
Cell Scraper	Sarstedt	831830	RT
Chloroform (zur Analyse)	AppliChem	A1585,0500	RT
Magnetic beads	Qiagen		RT
Ethanol (abs.) for Molecular Biology	AppliChem	A3678,100	RT

Preparation of Solutions and Buffers

Before using the RNEasy kit for the first time, prepare RPE buffer working solution by adding 4 volumes of ethanol (96%−100%) as indicated on the bottle.

Turn on the cooling centrifuge.

Procedure

1. Cut and weight the frozen piece of tissue (max. 30 mg) with a sterile scalpel.
2. Work in the fume hood: Transfer every piece of tissue to a RNAse-free Safe-Lock sterile Eppendorf tube containing 1 mL of TriPure and a magnetic bead.
3. Disrupt the tissue at the tissue lyser (first floor) at 25 Hz for 3 min (more if tissue not disrupted).
4. Let the samples rest for 5 min at RT.
5. Collect the lysate with a pipette and transfer to another RNAse-free Eppendorf tube.
6. Add 200 μL chloroform and shake it vigorously for 15 s.
7. Let the tubes rest for 2−3 min at RT.
8. Centrifuge at 12,000 ×g for 15 min at 4°C.
9. Transfer the upper aqueous phase (∼600 μL) to a new tube.*
   * (2 mL Eppendorf tube if the QIAcube is used; please see below)
10. Add 1 volume of 70% ethanol and mix thoroughly by vortexing (do not centrifuge).
11. Transfer up to 700 μL of the sample to an RNeasy Mini spin column placed in a 2 mL collection tube and centrifuge for 15 s at ≥8000 ×g (≥10,000 rpm) at RT; discard the flow-through.
12. Repeat step 10 using the remaining sample; discard the flow-through.
13. Add 700 μL buffer RW1 to the RNeasy spin column, centrifuge for 15 s at ≥8000 ×g (≥10,000 rpm), and discard the flow-through.
14. Add 500 μL buffer RPE to the RNeasy spin column, centrifuge for 15 s at ≥8000 ×g (≥10,000 rpm), and discard the flow-through.
15. Add 500 μL buffer RPE to the RNeasy spin column, centrifuge for 2 min at ≥8000 ×g (≥10,000 rpm), and discard the flow-through.

16. Place the RNeasy spin column in a new 2 mL collection tube and centrifuge at full speed for 1 min.
17. Place the RNeasy spin column in a new 1.5 mL collection tube, add 30−50 μL RNase-free water directly to the spin column membrane, and centrifuge for 1 min at $\geq 8000 \times g$ ($\geq 10,000$ rpm) to elute the RNA.
18. The RNA samples should be placed on ice after isolation.
19. Measure the RNA concentration using the NanoDrop. Measure in duplicates or triplicates using 1 μL of sample per measurement. Store RNA at $-20°C$.

*If the **QIAcube** is used, this protocol is followed; proceed until step 8. Use the centrifugation time to prepare the QIAcube:

1. Load the rotor adapters with RNeasy spin columns and 1.5 mL collection tubes and place them in the QIAcube centrifuge.
2. Place the bottles containing the buffers in the QIAcube, refill if necessary.
3. Load the tips.
4. Load the 2 mL Eppendorf tube containing the upper aqueous phase of the sample to the robot.
5. Adjust the elution volume if necessary.
6. Start the program: RNeasy Lipid—animal tissue—aqueous phase.

ISOLATION OF RNA FROM MOUSE WHITE ADIPOSE TISSUE

Materials
Consumables, buffer, media, and solutions

Materials	Distributor/ Manufacturer	Cat. No.	Storage
RNeasy Mini Kit (250)	Qiagen	74106	RT
TriPure	Roche	11667165001	4°C
Cell Scraper	Sarstedt	831830	RT
Chloroform	AppliChem	A1585,0500	RT
Magnetic beads	Qiagen		RT
Ethanol (abs.) for Molecular Biology	AppliChem	A3678,100	RT

Preparation of Solutions and Buffers
Before using the RNEasy kit for the first time, prepare RPE buffer working solution by adding 4 volumes of ethanol (96%−100%) as indicated on the bottle.
   Turn on the cooling centrifuge.

Procedure
1. Cut and weight the frozen piece of tissue (~ 100 mg) with a sterile scalpel.
2. Make sure to keep tissue frozen the whole time by using dry ice.
3. Transfer to a 2 mL RNAse-free Safe-Lock sterile Eppendorf tube.
4. Add 1 mL of TriPure and a magnetic bead.

5. Disrupt the tissue at the tissue lyser (first floor) at 25 Hz for 2×2 min (more if tissue not disrupted).
6. Let the samples rest for 5 min at RT.
7. Collect the lysate with a pipette and transfer to another RNAse-free Eppendorf tube.
8. Add 200 μL chloroform and shake it vigorously for 15 s.
9. Let the tubes rest for 2–3 min at RT.
10. Centrifuge at 12,000 ×g for 15 min at 4°C.
11. Transfer the upper aqueous phase (~600 μL) to a new tube.*
    * (2 mL Eppendorf tube if the QIAcube is used; please see below)
12. Add 1 volume of 70% ethanol and mix thoroughly by vortexing (do not centrifuge).
13. Transfer up to 700 μL of the sample to an RNeasy Mini spin column placed in a 2 mL collection tube and centrifuge for 15 s at ≥8000 ×g (≥10,000 rpm) at RT; discard the flow-through.
14. Add 700 μL buffer RW1 to the RNeasy spin column, centrifuge for 15 s at ≥8000 ×g (≥10,000 rpm), and discard the flow-through.
15. Add 500 μL buffer RPE to the RNeasy spin column, centrifuge for 15 s at ≥8000 ×g (≥10,000 rpm), and discard the flow-through.
16. Add 500 μL buffer RPE to the RNeasy spin column, centrifuge for 2 min at ≥8000 ×g (≥10,000 rpm), and discard the flow-through.
17. Place the RNeasy spin column in a new 2 mL collection tube and centrifuge at full speed for 1 min.
18. Place the RNeasy spin column in a new 1.5 mL collection tube, add 30 μL RNase-free water directly to the spin column membrane, and centrifuge for 1 min at ≥8000 ×g (≥10,000 rpm) to elute the RNA.
19. The RNA samples should be placed on ice after isolation.
20. Measure the RNA concentration using the NanoDrop. Measure in duplicates or triplicates using 1 μL of sample per measurement. Store RNA at −20°C.

RT-PCR USING GOTAQ QPCR MASTER MIX (PROMEGA)

Materials
Consumables, buffer, media, and solutions

Materials	Distributor/ Manufacturer	Cat. No.	Storage
GoTaq qPCR Master Mix (1.000)	Promega	A6002	−20°C
Fast Optical 96w plates barcoded	Applied Biosystems	4366932	RT
Optical adhesive covers	Applied Biosystems	4311971	RT
EDTA	Sigma	E5134	RT
Tris Ultrapure	AppliChem	A1086,1000	RT
H_2O for Molecular Biology	AppliChem	A7398,0500	RT

Preparation of Solutions and Buffers

1. Prepare TE-buffer (pH 8.0; contains 10 mM Tris·Cl and 1 mM EDTA).

   For 100 mL TE, mix the following stock solutions:
   - 1 mL of 1 M Tris·Cl, pH 8.0 (autoclaved)
   - 0.2 mL of 0.5 M EDTA, pH 8.0 (autoclaved)
   - 98.8 mL of distilled H_2O

2. Dissolve the primer in TE-buffer according to the instruction of the supplier.

Procedure

1. Thaw all reagents and the cDNA and place them on ice.
2. Prepare a master mix for each primer. Each sample is analyzed in triplicates, and an H_2O sample is used as negative control. Per well you need the following:

Using Qiagen Quantitect Primer:	**OR** Using Eurofins MWG Primer:
5 µL GoTaq qPCR Master Mix	5 µL GoTaq qPCR Master Mix
1 µL QuantiTect Primer (Qiagen)	1 µL forward Primer (Eurofins MWG)
2.9 µL H_2O	1 µL reverse Primer (Eurofins MWG)
0.1 LI CXR Dye	1.9 µL H_2O
	0.1 LI CXR Dye
9 µL total volume	9 µL total volume

3. Vortex thoroughly and centrifuge briefly.
4. Pipette 9 µL mastermix per well according to the plate layout (this can be carried out at RT).
5. Add 1 µL cDNA template or H_2O (neg. control) per well.
6. Seal the plate, vortex, and centrifuge briefly.
7. Run your program on the Cycler (StepOne Plus System).

STANDARD PROGRAM FOR QUANTITECT PRIMER ASSAY

Initiation:	2 min at 95°C	
Cycling Stage:	15 s at 95°C	
	30 s at 55°C	
	30 s at 60°C	40 Cycles
Followed by melting curve analysis		

REFERENCES

A collection of references regarding work published by the author related to secretome analysis and cellular crosstalk is presented on the following pages. The methods described in this chapter have been used successfully in these publications.

Secretome of human adipocytes

Dietze, D., Koenen, M., Rohrig, K., Horikoshi, H., Hauner, H., Eckel, J., 2002. Impairment of insulin signaling in human skeletal muscle cells by co-culture with human adipocytes. Diabetes 51, 2369−2376.

Famulla, S., Horrighs, A., Cramer, A., Sell, H., Eckel, J., 2012. Hypoxia reduces the response of human adipocytes towards TNFalpha resulting in reduced NF-kappaB signaling and MCP-1 secretion. Int. J. Obes. 36, 986−992.

Famulla, S., Lamers, D., Hartwig, S., Passlack, W., Horrighs, A., Cramer, A., Lehr, S., Sell, H., Eckel, J., 2011. Pigment epithelium-derived factor (PEDF) is one of the most abundant proteins secreted by human adipocytes and induces insulin resistance and inflammatory signaling in muscle and fat cells. Int. J. Obes. 35, 762−772.

Famulla, S., Schlich, R., Sell, H., Eckel, J., 2012. Differentiation of human adipocytes at physiological oxygen levels results in increased adiponectin secretion and isoproterenol-stimulated lipolysis. Adipocyte 1, 132−181.

Fleckenstein-Elsen, M., Dinnies, D., Jelenik, T., Roden, M., Romacho, T., Eckel, J., 2016. Eicosapentaenoic acid and arachidonic acid differentially regulate adipogenesis, acquisition of a brite phenotype and mitochondrial function in primary human adipocytes. Mol. Nutr. Food Res.

Lehr, S., Hartwig, S., Lamers, D., Famulla, S., Muller, S., Hanisch, F.G., Cuvelier, C., Ruige, J., Eckardt, K., Ouwens, D.M., Sell, H., Eckel, J., 2012. Identification and validation of novel adipokines released from primary human adipocytes. Mol. Cell. Proteomics 11. M111 010504.

Liu, L.S., Spelleken, M., Rohrig, K., Hauner, H., Eckel, J., 1998. Tumor necrosis factor-alpha acutely inhibits insulin signaling in human adipocytes: implication of the p80 tumor necrosis factor receptor. Diabetes 47, 515−522.

Muller, G., Wied, S., Crecelius, A., Kessler, A., Eckel, J., 1997. Phosphoinositolglycan-peptides from yeast potently induce metabolic insulin actions in isolated rat adipocytes, cardiomyocytes, and diaphragms. Endocrinology 138, 3459−3475.

Rohrborn, D., Bruckner, J., Sell, H., Eckel, J., 2016. Reduced DPP4 activity improves insulin signaling in primary human adipocytes. Biochem. Biophys. Res. Commun. 471, 348−354.

Rohrborn, D., Eckel, J., Sell, H., 2014. Shedding of dipeptidyl peptidase 4 is mediated by metalloproteases and up-regulated by hypoxia in human adipocytes and smooth muscle cells. FEBS Lett. 588, 3870−3877.

Romacho, T., Glosse, P., Richter, I., Elsen, M., Schoemaker, M.H., van Tol, E.A., Eckel, J., 2015. Nutritional ingredients modulate adipokine secretion and inflammation in human primary adipocytes. Nutrients 7, 865−886.

Schlich, R., Willems, M., Greulich, S., Ruppe, F., Knoefel, W.T., Ouwens, D.M., Maxhera, B., Lichtenberg, A., Eckel, J., Sell, H., 2013. VEGF in the crosstalk between human adipocytes and smooth muscle cells: depot-specific release from visceral and perivascular adipose tissue. Mediat. Inflamm. 2013, 982458.

Sell, H., Dietze-Schroeder, D., Eckardt, K., Eckel, J., 2006. Cytokine secretion by human adipocytes is differentially regulated by adiponectin, AICAR, and troglitazone. Biochem. Biophys. Res. Commun. 343, 700−706.

Sell, H., Eckel, J., 2007. Regulation of retinol binding protein 4 production in primary human adipocytes by adiponectin, troglitazone and TNF-alpha. Diabetologia 50, 2221–2223.

Tomazic, M., Janez, A., Sketelj, A., Kocijancic, A., Eckel, J., Sharma, P.M., 2002. Comparison of alterations in insulin signalling pathway in adipocytes from Type II diabetic pregnant women and women with gestational diabetes mellitus. Diabetologia 45, 502–508.

Zhou, L., Sell, H., Eckardt, K., Yang, Z., Eckel, J., 2007. Conditioned medium obtained from in vitro differentiated adipocytes and resistin induce insulin resistance in human hepatocytes. FEBS Lett. 581, 4303–4308.

Myokines and skeletal muscle

Bahr, M., Kolter, T., Seipke, G., Eckel, J., 1997. Growth promoting and metabolic activity of the human insulin analogue [GlyA21,ArgB31,ArgB32]insulin (HOE 901) in muscle cells. Eur. J. Pharmacol. 320, 259–265.

Dietze, D., Koenen, M., Rohrig, K., Horikoshi, H., Hauner, H., Eckel, J., 2002. Impairment of insulin signaling in human skeletal muscle cells by co-culture with human adipocytes. Diabetes 51, 2369–2376.

Dietze, D., Ramrath, S., Ritzeler, O., Tennagels, N., Hauner, H., Eckel, J., 2004. Inhibitor kappaB kinase is involved in the paracrine crosstalk between human fat and muscle cells. Int. J. Obes. Relat. Metab. Disord. 28, 985–992.

Eckardt, K., May, C., Koenen, M., Eckel, J., 2007. IGF-1 receptor signalling determines the mitogenic potency of insulin analogues in human smooth muscle cells and fibroblasts. Diabetologia 50, 2534–2543.

Eckardt, K., Schober, A., Platzbecker, B., Mracek, T., Bing, C., Trayhurn, P., Eckel, J., 2011. The adipokine zinc-alpha2-glycoprotein activates AMP kinase in human primary skeletal muscle cells. Arch. Physiol. Biochem. 117, 88–93.

Eckardt, K., Sell, H., Eckel, J., 2008. Novel aspects of adipocyte-induced skeletal muscle insulin resistance. Arch. Physiol. Biochem. 114, 287–298.

Eckardt, K., Sell, H., Taube, A., Koenen, M., Platzbecker, B., Cramer, A., Horrighs, A., Lehtonen, M., Tennagels, N., Eckel, J., 2009. Cannabinoid type 1 receptors in human skeletal muscle cells participate in the negative crosstalk between fat and muscle. Diabetologia 52, 664–674.

Famulla, S., Lamers, D., Hartwig, S., Passlack, W., Horrighs, A., Cramer, A., Lehr, S., Sell, H., Eckel, J., 2011. Pigment epithelium-derived factor (PEDF) is one of the most abundant proteins secreted by human adipocytes and induces insulin resistance and inflammatory signaling in muscle and fat cells. Int. J. Obes. 35, 762–772.

Gorgens, S.W., Eckardt, K., Elsen, M., Tennagels, N., Eckel, J., 2014. Chitinase-3-like protein 1 protects skeletal muscle from TNFalpha-induced inflammation and insulin resistance. Biochem. J. 459, 479–488.

Gorgens, S.W., Raschke, S., Holven, K.B., Jensen, J., Eckardt, K., Eckel, J., 2013. Regulation of follistatin-like protein 1 expression and secretion in primary human skeletal muscle cells. Arch. Physiol. Biochem. 119, 75–80.

Hartwig, S., Raschke, S., Knebel, B., Scheler, M., Irmler, M., Passlack, W., Muller, S., Hanisch, F.G., Franz, T., Li, X., Dicken, H.D., Eckardt, K., Beckers, J., de Angelis, M.H., Weigert, C., Haring, H.U., Al-Hasani, H., Ouwens, D.M., Eckel, J., Kotzka, J., Lehr, S., 2014. Secretome profiling of primary human skeletal muscle cells. Biochim. Biophys. Acta 1844, 1011–1017.

Indrakusuma, I., Sell, H., Eckel, J., 2015. Novel mediators of adipose tissue and muscle crosstalk. Curr Obes Rep 4, 411−417.

Kanzleiter, T., Rath, M., Gorgens, S.W., Jensen, J., Tangen, D.S., Kolnes, A.J., Kolnes, K.J., Lee, S., Eckel, J., Schurmann, A., Eckardt, K., 2014. The myokine decorin is regulated by contraction and involved in muscle hypertrophy. Biochem. Biophys. Res. Commun. 450, 1089−1094.

Kessler, A., Muller, G., Wied, S., Crecelius, A., Eckel, J., 1998. Signalling pathways of an insulin-mimetic phosphoinositolglycan-peptide in muscle and adipose tissue. Biochem. J. 330 (Pt 1), 277−286.

Lambernd, S., Taube, A., Schober, A., Platzbecker, B., Gorgens, S.W., Schlich, R., Jeruschke, K., Weiss, J., Eckardt, K., Eckel, J., 2012. Contractile activity of human skeletal muscle cells prevents insulin resistance by inhibiting pro-inflammatory signalling pathways. Diabetologia 55, 1128−1139.

Raschke, S., Eckardt, K., Bjorklund Holven, K., Jensen, J., Eckel, J., 2013. Identification and validation of novel contraction-regulated myokines released from primary human skeletal muscle cells. PLoS One 8, e62008.

Schreyer, S., Ledwig, D., Rakatzi, I., Kloting, I., Eckel, J., 2003. Insulin receptor substrate-4 is expressed in muscle tissue without acting as a substrate for the insulin receptor. Endocrinology 144, 1211−1218.

Sell, H., Dietze-Schroeder, D., Kaiser, U., Eckel, J., 2006. Monocyte chemotactic protein-1 is a potential player in the negative cross-talk between adipose tissue and skeletal muscle. Endocrinology 147, 2458−2467.

Sell, H., Eckardt, K., Taube, A., Tews, D., Gurgui, M., Van Echten-Deckert, G., Eckel, J., 2008. Skeletal muscle insulin resistance induced by adipocyte-conditioned medium: underlying mechanisms and reversibility. Am. J. Physiol. Endocrinol. Metab. 294, E1070−E1077.

Sell, H., Eckel, J., Dietze-Schroeder, D., 2006. Pathways leading to muscle insulin resistance−the muscle−fat connection. Arch. Physiol. Biochem. 112, 105−113.

Sell, H., Jensen, J., Eckel, J., 2012. Measurement of insulin sensitivity in skeletal muscle in vitro. Meth. Mol. Biol. 933, 255−263.

Sell, H., Kaiser, U., Eckel, J., 2007. Expression of chemokine receptors in insulin-resistant human skeletal muscle cells. Horm. Metab. Res. 39, 244−249.

Sell, H., Laurencikiene, J., Taube, A., Eckardt, K., Cramer, A., Horrighs, A., Arner, P., Eckel, J., 2009. Chemerin is a novel adipocyte-derived factor inducing insulin resistance in primary human skeletal muscle cells. Diabetes 58, 2731−2740.

Taube, A., Lambernd, S., van Echten-Deckert, G., Eckardt, K., Eckel, J., 2012. Adipokines promote lipotoxicity in human skeletal muscle cells. Arch. Physiol. Biochem. 118, 92−101.

Trayhurn, P., Drevon, C.A., Eckel, J., 2011. Secreted proteins from adipose tissue and skeletal muscle - adipokines, myokines and adipose/muscle cross-talk. Arch. Physiol. Biochem. 117, 47−56.

Vahsen, S., Rakowski, K., Ledwig, D., Dietze-Schroeder, D., Swifka, J., Sasson, S., Eckel, J., 2006. Altered GLUT4 translocation in skeletal muscle of 12/15-lipoxygenase knockout mice. Horm. Metab. Res. 38, 391−396.

Vascular cells

Eckardt, K., May, C., Koenen, M., Eckel, J., 2007. IGF-1 receptor signalling determines the mitogenic potency of insulin analogues in human smooth muscle cells and fibroblasts. Diabetologia 50, 2534−2543.

Lamers, D., Schlich, R., Greulich, S., Sasson, S., Sell, H., Eckel, J., 2011. Oleic acid and adipokines synergize in inducing proliferation and inflammatory signalling in human vascular smooth muscle cells. J. Cell Mol. Med. 15, 1177−1188.

Lamers, D., Schlich, R., Horrighs, A., Cramer, A., Sell, H., Eckel, J., 2012. Differential impact of oleate, palmitate, and adipokines on expression of NF-kappaB target genes in human vascular smooth muscle cells. Mol. Cell. Endocrinol. 362, 194−201.

Rohrborn, D., Eckel, J., Sell, H., 2014. Shedding of dipeptidyl peptidase 4 is mediated by metalloproteases and up-regulated by hypoxia in human adipocytes and smooth muscle cells. FEBS Lett. 588, 3870−3877.

Sasson, S., Eckel, J., 2006. Disparate effects of 12-lipoxygenase and 12-hydroxyeicosatetraenoic acid in vascular endothelial and smooth muscle cells and in cardiomyocytes. Arch. Physiol. Biochem. 112, 119−129.

Schlich, R., Lamers, D., Eckel, J., Sell, H., 2015. Adipokines enhance oleic acid-induced proliferation of vascular smooth muscle cells by inducing CD36 expression. Arch. Physiol. Biochem. 121, 81−87.

Schlich, R., Willems, M., Greulich, S., Ruppe, F., Knoefel, W.T., Ouwens, D.M., Maxhera, B., Lichtenberg, A., Eckel, J., Sell, H., 2013. VEGF in the crosstalk between human adipocytes and smooth muscle cells: depot-specific release from visceral and perivascular adipose tissue. Mediat. Inflamm. 2013, 982458.

Wronkowitz, N., Gorgens, S.W., Romacho, T., Villalobos, L.A., Sanchez-Ferrer, C.F., Peiro, C., Sell, H., Eckel, J., 2014. Soluble DPP4 induces inflammation and proliferation of human smooth muscle cells via protease-activated receptor 2. Biochim. Biophys. Acta 1842, 1613−1621.

Index

Printed in the United States
By Bookmasters